Fuzzy Transforms for Image Processing and Data Analysis

Ferdinando Di Martino · Salvatore Sessa

Fuzzy Transforms for Image Processing and Data Analysis

Core Concepts, Processes and Applications

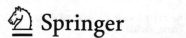 Springer

Ferdinando Di Martino ⓘ
Department of Architecture
University of Naples Federico II
Naples, Italy

Salvatore Sessa ⓘ
Department of Architecture
University of Naples Federico II
Naples, Italy

ISBN 978-3-030-44615-4   ISBN 978-3-030-44613-0   (eBook)
https://doi.org/10.1007/978-3-030-44613-0

This Springer imprint is published by the registered company Springer Nature Switzerland AG
The registered company address is: Gewerbestrasse 11, 6330 Cham, Switzerland

Preface

Since the publication of the paper by Prof. Irina Perfilieva in 2004, Fuzzy Transforms were applied in a wide range of fields.

Fuzzy Transform is a function approximation method based on fuzzy sets, similar to the Fourier Transform.

This book is devoted to researchers, students, and professionals who intend to explore the use of techniques based on fuzzy transforms in image and data analysis. In what follows, our attention is focused primarily on the applicability of this technique in specific issues such as image and video compression, image segmentation, forecasting, and data classification.

Applications in which techniques based on the use of the Fuzzy Transform are used to deal with these problems are discussed in detail and the algorithms that implement them are formally shown in the pseudocode. Of course, these issues do not exhaust the whole range of problems in which the fuzzy transform technique can be successfully applied; the reader may eventually deepen the study and analysis of the wide range of techniques based on fuzzy transforms used in the various application contexts.

Chapter 1 defines some of the basic notations within the conceptual framework of fuzzy transform where the concepts of uniform and h-uniform fuzzy partition and direct and inverse fuzzy transform are analyzed as well. In Chap. 2, two extensions of the fuzzy transform are presented: the multi-dimensional fuzzy transform and the high degree fuzzy transform.

Chapter 3 deals with the use of the two-dimensional fuzzy transform for compressing images and videos and in Chap. 4 an application for the passive image autofocus is discussed.

Chapter 5 focuses on the use of techniques based on fuzzy transform and F^1-transform in image fusion and edge detection. Chapter 6 deals with methods based on fuzzy transform in image segmentation and image thresholding. In Chap. 7 two methods based on fuzzy transform are explored for image watermarking and tamper detection.

The use of fuzzy transform in data analysis is introduced in Chap. 8 precisely two methods analyze attribute dependency and mining association rules.

Chapter 9 deals with techniques concerning forecasting and time series analysis and, in particular, in Chap. 10 techniques based on fuzzy transform and F^1-transform are explored for seasonal time series analysis.

Chapter 11 analyzes machine learning methods based on multi-dimensional fuzzy transforms for data classification. Chapter 12 explores the use of techniques for the analysis of massive data: in particular, a fuzzy transform-based method for analyzing the dependency between attributes in massive datasets is discussed.

The relevant role that today the research of machine learning techniques and innovative methods to process and analyze massive data is certainly a spur for researchers to be implemented in the future, ever more algorithms based on fuzzy transform and hybrids are expected as well.

We hope that other researchers, inspired from this book, contribute in these years to the fast growing of fuzzy transforms theory and related applications.

We also thank Ph.D. Barbara Cardone for her constant help in drafting and correcting this book.

Naples, Italy   Ferdinando Di Martino
                                                                       Salvatore Sessa

Contents

Chapter 1
Fuzzy Transform Concepts

In this chapter we introduce the basic concepts of fuzzy transform (F-transform, for short). In the first paragraph, the basics concerning fuzzy sets and fuzzy relationships are discussed. Then the concepts of fuzzy partition, uniform and h-uniform fuzzy partitions are introduced. The direct and inverse F-transform definitions are given and the discrete F-transform concept is presented.

1.1 Fuzzy Sets and Fuzzy Relations

Fuzzy set theory was introduced by Zadeh [8] as a generalization of classical set theory in order to manage problems in which information is incomplete or imprecise.

Let U be a collection of objects called *universe of discourse*. Then, a *fuzzy set A* in X is a set of ordered pairs: $A = \{(x, \mu_A(x)) \in U\}$ where $\mu_A \colon U \to [0,1]$ is called the *membership function* or *membership degree* of x in A.

If $\mu_A(x) \in \{0,1\}$ for every $x \in U$, A is reduced to a classical set, called *crisp set*, and given by a subset of U.

Given a value $\alpha \in [0, 1]$, we denote with A_α, called *α-cut* of the fuzzy set A, the crisp set containing the elements $x \in U$ with a membership degree greater or equal to α.

$$A_\alpha = \{x \in U \mid \mu_A(x) \geq \alpha\} \tag{1.1}$$

We denote with A'_α, called *strong α-cut* of the fuzzy set A, the crisp set containing the elements $x \in U$ with a membership degree greater than α.

$$A'_\alpha = \{x \in U \mid \mu_A(x) > \alpha\} \tag{1.2}$$

The *α-cut of A with α* = 1, A_1, is called the *core* of the fuzzy set A and denoted by $C(A)$:

© Springer Nature Switzerland AG 2020
F. Di Martino and S. Sessa, *Fuzzy Transforms for Image Processing and Data Analysis*,
https://doi.org/10.1007/978-3-030-44613-0_1

$$C(A) = \{x \in U | \mu_A(x) = 1\} \tag{1.3}$$

We call support of the fuzzy set A, and denote it by $S(A)$, the crisp set A_0' given by the strong α-*cut of A with* $\alpha = 0$:

$$S(A) = \{x \in U | \mu_A(x) > 0\} \tag{1.4}$$

The *height* of a fuzzy set, denoted by Hgt(A) is given by the max membership degree

$$\mathrm{Hgt}(A) = \sup_{x \in U} \mu_A(x) \tag{1.5}$$

A fuzzy set is said to be *normalized* iff Hgt(A) $= 1$.

If the universe of discourse is a subset of the real line ($U \subseteq \mathbb{R}$), its elements x are real numbers and we can define the *width* of the fuzzy set A, denoted by Width(A), given by

$$\mathrm{Width}(A) = \max(S(A)) - \min(S(A)) \tag{1.6}$$

A real fuzzy set A is said to be *convex* if $\forall x, y \in U, \forall \lambda \in [0, 1] : \mu_A(\lambda x + (1 - \lambda)y) \geq \min(\mu_A(x), \mu_A(y))$.

Many types of membership functions are used in the literature; generally simple functions are used to build fuzzy sets because to model fuzzy concepts the use of more complex functions does not add more precision.

Now we report the best-known fuzzy sets.

A *trapezoidal* fuzzy set is constructed by assigning four parameters on the universe of the discourse (say) a, b, c, d, where $a \leq c \leq d \leq b$.

$$\text{Trapezoidal } \mu_A(x) = \begin{cases} 0 & x < a \\ \frac{(x-a)}{c-a} & a \leq x \leq c \\ 1 & c \leq x \leq d \\ \frac{b-x}{b-d} & d \leq x \leq b \\ 0 & x > b \end{cases} \tag{1.7}$$

When $c = d$ we obtain a *triangular* fuzzy set:

$$\text{Triangular } \mu_A(x) = \begin{cases} 0 & x < a \\ \frac{(x-a)}{c-a} & a \leq x \leq c \\ \frac{b-x}{b-c} & c \leq x \leq b \\ 0 & x > b \end{cases} \tag{1.8}$$

If $a = c = -\infty$ we obtain an *R-function* fuzzy set, given by

$$R - \text{function } \mu_A(x) = \begin{cases} 1 & x < d \\ \frac{(x-d)}{b-d} & d \le x \le b \\ 0 & x > b \end{cases} \qquad (1.9)$$

And if $d = b = +\infty$ we obtain an *L-function* fuzzy set, given by

$$L - \text{function } \mu_A(x) = \begin{cases} 0 & x < a \\ \frac{(x-c)}{c-a} & a \le x \le c \\ 1 & x > c \end{cases} \qquad (1.10)$$

When $a = b = c = d$ we obtain a *Gaussian* fuzzy set constructed by assigning two parameters, m and s, with $s > 0$, representing the mean and the standard deviation of a Gaussian distribution. We obtain

$$\text{Gaussian } \mu_A(x) = e^{-\frac{(x-m)^2}{2s^2}} \qquad (1.11)$$

A *sigmoid* fuzzy set is constructed assigning two parameters, α and β. The dimensionless parameter α gives the steepness of the curve, β is the sigmoid midpoint. The sigmoid is increasing if $\alpha > 0$, decreasing if $\alpha < 0$. We obtain

$$\text{Sigmoid } \mu_A(x) = \frac{1}{1 + e^{-\alpha(x-\beta)}} \qquad (1.12)$$

Below we show examples of fuzzy sets constructed on a continuous domain given by the air temperature in Celsius degree. The fuzzy set labeled *Cold* is constructed with an R-function ($d = 5$ °C, $b = 15$ °C), the fuzzy set *Mean* is constructed with three different membership functions: trapezoidal ($a = 8$ °C, $c = 16$ °C, $d = 20$ °C, $b = 27$ °C), triangular ($a = 8$ °C, $c = 18$ °C, $b = 27$ °C), and a Gaussian function ($m = 18$ °C, $s = 4$ °C), the fuzzy set *Cold* is constructed by using both a sigmoidal ($\alpha = 1$, $\beta = 30$ °C) and an L-function ($a = 20$ °C, $c = 32$ °C) (Fig. 1.1).

The *complement* of the fuzzy set A, denoted by A^C, is given by the fuzzy set:

$$A^C = \{(x, 1 - \mu_A(x)) | x \in U\} \qquad (1.13)$$

The *t-norm intersection* of two fuzzy sets A and B is defined as

$$A \cap_t B = \{(x, t(\mu_A(x), \mu_B(x)) | x \in U\} \qquad (1.14)$$

where the function $t: [0, 1] \times [0, 1] \to [0, 1]$ is called *triangular norm* or *t-norm*. A *t*-norm satisfies the following properties:

1. Commutativity: $\forall a, b \in [0,1]: t(a, b) = t(b, a)$;
2. Monotonicity: $\forall a, b, c, d \in [0,1]$ with $a \le c$ and $b \le d$: $t(a, b) \le t(c, d)$;
3. Associativity: $\forall a, b, c \in [0,1]: t(a, t(b, c)) = t(t(a, b), c)$;
4. Existence of the identity element 1: $\forall a \in [0,1]: t(a, 1) = t(1, a) = a$.

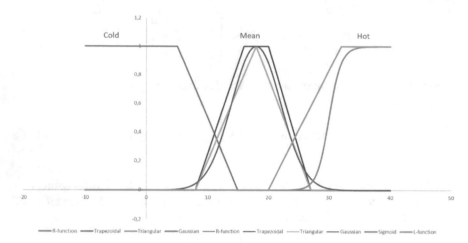

Fig. 1.1 Examples of fuzzy sets on a continuous domain

A complete analysis of the various families of t-norms is performed in [4]. Well-known examples of t-norms used in fuzzy logic are

- *Minimum t-norm*—$t(a, b) = \min(a, b)$—also called Gödel t-norm, as it is the standard semantics for conjunction in Gödel fuzzy logic. Besides that, it occurs in most t-norm-based fuzzy logics as the standard semantics for weak conjunction.
- *Product t-norm*—$t(a, b)) = a \cdot b$. Besides other uses, the product t-norm is the standard semantics for strong conjunction in product fuzzy logic. It is a strict Archimedean t-norm.
- *Łukasiewicz t-norm*—$t(a, b) = \max(0, a + b - 1)$. The name comes from the fact that the t-norm is the standard semantics for strong conjunction in Łukasiewicz fuzzy logic. It is a nilpotent Archimedean t-norm, pointwise smaller than the product t-norm.

In Figs. 1.2, 1.3, and 1.4, we show the Min, Product, and Lukasiewicz t-norms between a trapezoidal and a Gaussian fuzzy sets, respectively.

The *s-union* of two fuzzy sets A and B is defined as

$$A \cup_s B = \{(x, s(\mu_A(x), \mu_B(x)) | x \in U\} \tag{1.15}$$

where the function $s: [0, 1] \times [0, 1] \rightarrow [0, 1]$ is called *triangular conorm*, or *s-norm*. An s-norm satisfies the following properties:

1. Commutativity: $\forall a, b \in [0,1]$: $s(a, b) = s(b, a)$;
2. Monotonicity: $\forall a, b, c, d \in [0,1]$ with $a \leq c$ and $b \leq d$: $t(a, b) \leq t(c, d)$;
3. Associativity: $\forall a, b, c \in [0,1]$: $s(a, s(b, c)) = s(s(a, b), c)$;
4. Existence of the null element 0: $\forall a \in [0,1]$: $s(a, 0) = s(0, a) = a$.

T-norm and s-norm are logical duals and we can obtain the corresponding s-norm by a t-norm, with the formula:

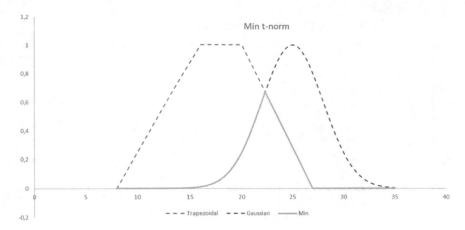

Fig. 1.2 Example of min *t*-norm

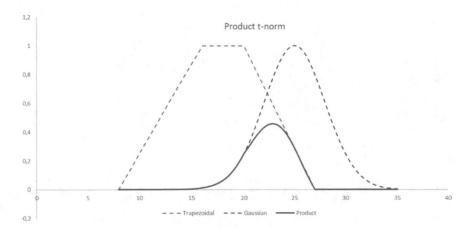

Fig. 1.3 Example of product *t*-norm

$$s(a, b) = 1 - t(1 - a, 1 - b) \tag{1.16}$$

Well-known examples of *s*-norms used in fuzzy logic are

- *Maximum s-norm*—$s(a, b) = \max(a, b)$—also called Gödel *s*-norm.
- *Algebraic sum s-norm*—$s(a, b)) = a + b - a \cdot b$—also called *probabilistic sum*.
- *Łukasiewicz s-norm*—$s(a, b) = \min(a + b, 1)$—also called *bounded sum*.

In Figs. 1.5, 1.6, and 1.7 are shown, respectively, the Min, Product, and Lukasiewicz *s*-norms between a trapezoidal and a Gaussian fuzzy sets.

The *t-difference* of two fuzzy sets A and B is defined as

$$A \backslash_t B = A \cap_t B^C = \{(x, t(\mu_A(x), 1 - \mu_B(x)) | x \in U\} \tag{1.17}$$

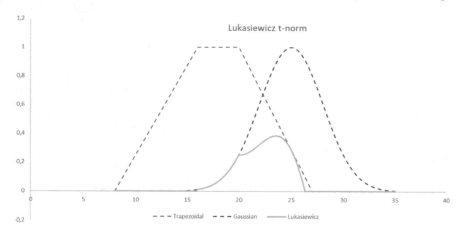

Fig. 1.4 Example of Lukasiewicz *t*-norm

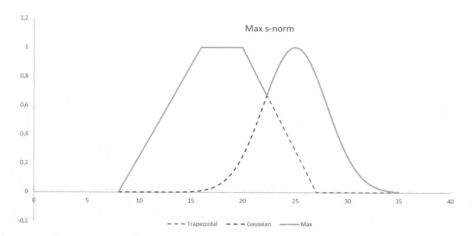

Fig. 1.5 Example of min *s*-norm

Given a positive integer *n*, the *n-power* of the fuzzy set *A* is the fuzzy set given by

$$A^n = \left\{ (x, \mu_A(x)^n) | x \in U \right\} \tag{1.18}$$

The 2-power of *A*, A^2 is called concentration of *A*, denoted also as Con(*A*).

Let *X* and *Y* be two universal sets. A fuzzy relation *R* is a subset of the Cartesian product $X \times Y$ [2] describing a map $R: X \times Y \to X \times Y$ and given by

$$R = \{((x, y), \mu_R(x, y)) | (x, y) \in X \times Y\} \tag{1.19}$$

where $\mu_R(x, y) \in [0, 1]$ for every $(x, y) \in X \times Y$.

Fuzzy relations are generalizations of crisp relations, in which $\mu_R(x, y) \in \{0, 1\}$.

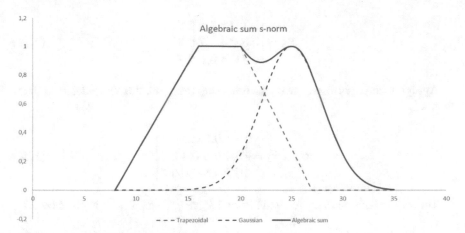

Fig. 1.6 Example of algebraic sum s-norm

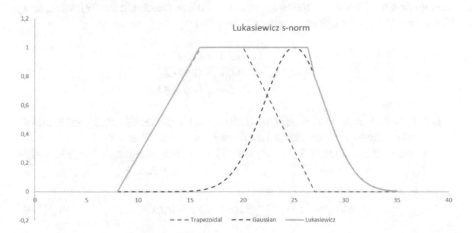

Fig. 1.7 Example of Lukasiewicz s-norm

The t-intersection of two fuzzy relations R_1 and R_2 defined on $X \times Y$ *is* given by

$$R_1 \cap_t R_2 = \left\{ \left((x, y), t(\mu_{R_1}(x, y), \mu_{R_2}(x, y)) \right) | (x, y) \in X \times Y \right\}. \quad (1.20)$$

For example, let $X = \{x_1, x_2, x_3\}$ and $Y = \{y_1, y_2, y_3, y_4\}$ and consider the following fuzzy relations: $R_1 = $ "x is colder than y," $R_2 = $ "x is rainier than y," where

$$R_1 = \begin{vmatrix} 0.1 & 0.2 & 0.3 & 0.6 \\ 0.4 & 0.6 & 0.4 & 0.7 \\ 0.5 & 0.7 & 0.6 & 0.8 \end{vmatrix}$$

$$R_2 = \begin{vmatrix} 0.1\ 0.2\ 0.5\ 0.9 \\ 0.6\ 0.4\ 0.7\ 0.8 \\ 0.5\ 0.5\ 0.6\ 0.7 \end{vmatrix} \qquad (1.21)$$

Applying min t-norm, we obtain the following fuzzy relation "x colder and rainier than y."

$$R_1 \cap_t R_2 = \begin{vmatrix} 0.1\ 0.2\ 0.3\ 0.6 \\ 0.4\ 0.4\ 0.4\ 0.7 \\ 0.5\ 0.5\ 0.6\ 0.7 \end{vmatrix} \qquad (1.22)$$

The *s-union* of two fuzzy relations R_1 and R_2 defined on $X \times Y$ is given by

$$R_1 \cup_s R_2 = \left\{ ((x, y), s(\mu_{R_1}(x, y), \mu_{R_2}(x, y))) | (x, y) \in X \times Y \right\} \qquad (1.23)$$

Applying max s-norm to the two previous relations we obtain the following fuzzy relation $R_1\ U\ R_2 = $ "x colder or rainier than y."

$$R_1 \cup_s R_2 = \begin{vmatrix} 0.2\ 0.2\ 0.5\ 0.9 \\ 0.6\ 0.6\ 0.4\ 0.8 \\ 0.5\ 0.7\ 0.6\ 0.8 \end{vmatrix} \qquad (1.24)$$

Let $R_1 \subseteq X \times Y$ be a fuzzy relation defined on the universal sets X and Y and $R_2 \subseteq Y \times Z$ be a fuzzy relation defined on the universal sets Y and Z.

The fuzzy composition of R_1 and R_2 denoted as $R_1 \circ R_2$ is expressed by the relation [1]:

$$R_1 \circ R_2 = \left\{ \left((x, z), \bigvee_y \{\mu_{R_1}(x, y) \wedge \mu_{R_2}(y, z)\} \right) \right\} \qquad (1.25)$$

with $(x, y) \in X \times Y, (y, z) \in Y \times Z$.

For brevity, if the symbol $\bigwedge$ (resp., $\bigvee$) denotes the min t-norm (resp., max s-norm), then (1.22) is called max–min *composition* of R_1 and R_2.

1.2   Generalized Fuzzy Partitions and Fuzzy Partitions Under Ruspini Condition

Let X be a crisp set. A *partition* of X is a set of nonempty subsets of X such that every element x in X is in exactly one of these subsets. This concept can be generalized in fuzzy set theory. We refer to the concept of Fuzzy Partition introduced in [7] as the fuzzy sets $\{A_1, A_2, \ldots, A_n\}$ if

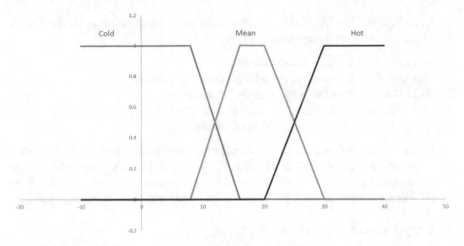

Fig. 1.8 Example of a Ruspini fuzzy partition

$$A_k(x) \neq 0 \quad \forall k = 1, \ldots, n \tag{1.26}$$

$$\sum_{k=1}^{n} A_k(x) = 1 \quad \forall x \in X \tag{1.27}$$

(1.27) is called the *Ruspini condition*.

As an example, we consider the following fuzzy partition $\{A_1, A_2, A_3\}$ of a crisp given X by the daily temperature, where

- A_1 is a fuzzy set labeled *Cold* and constructed with an R-function ($d = 8\ °C, b = 16\ °C$);
- A_2 is a fuzzy set labeled *Mean* and constructed with a trapezoidal function ($a = 8\ °C, c = 16\ °C, c = 20\ °C, c = 30\ °C$);
- A_3 is a fuzzy set labeled *Hot* and constructed with an L-function ($a = 20\ °C, b = 30\ °C$).

$\{A_1, A_2, A_3\}$ is a fuzzy partition of X because (1.24) holds for each x in X. In Fig. 1.8 we show this fuzzy partition.

1.3   Uniform Fuzzy Partition and H-Uniform Generalized Fuzzy Partition

Let $X = [a, b]$ be a closed interval of R; in [5, 6] was introduced the following definition of the fuzzy partition of X.

Let $x_0, x_1, x_2, \ldots, x_n$ be a set of $n + 1$ fixed points, called *nodes*, in $[a, b]$ such that $n \geq 3$ and $a = x_0 < x_1 < x_2 < \cdots < x_n = b$. We say that fuzzy sets

$A_1, \ldots, A_n : [a, b] \rightarrow [0, 1]$ form a *generalized fuzzy partition* of $[a, b]$, if for each $k = 1, 2, \ldots, n$ the following constraints hold:

1. $A_k(x) = 0 \ \forall x \notin (x_{k-1}, x_{k+1})$ locality;
2. $A_k(x) > 0 \ \forall x \in (x_{k-1}, x_{k+1})$ and $A_k(x_k) = 1$ positivity;
3. A_k is continuous in $[x_k - h'_k, x_k + h''_k]$ (continuity);
4. A_k is strictly decreasing in (x_{k-1}, x_k) and strictly increasing in (x_k, x_{k+1});
5. $\sum_{k=1}^{n} A_k(x) = 1 \ \forall x \in [a, b]$ (Ruspini condition).

The membership functions $\{A_1, \ldots, A_n\}$ are called *basic functions*. If the nodes $x_1, \ldots, x_n$ are equidistant, the fuzzy partition $\{A_1, \ldots, A_n\}$ is called h-uniform fuzzy partition of $[a, b]$. where $h = (b-a)/(n + 1)$ is the distance between two consecutive nodes. For an h-uniform fuzzy partition the following additional properties hold:

6. $A_k(x_k - x) = A_k(x_k + x) \ \forall x \in [0, h]$

7. $A_k(x) = A_{k-1}(x - h)$ and $A_{k-1}(x) = A_k(x + h) \ \forall x \in [x_k, x_{k+1}]$

An h-uniform fuzzy partition can be generated [3] by an even function $A_0 : [-1, 1] \rightarrow [0, 1]$, which is continuous, positive in $(-1, 1)$, and null on the boundaries $\{-1, 1\}$. The function A_0 is called *generating function* of the h-uniform fuzzy partition. The following expression represents an arbitrary h-uniform generalized fuzzy partition:

$$A_k(t) = \begin{cases} A_0\left(\frac{x - x_k}{h}\right) & x \in [x_k - h, x_k + h] \\ 0 & \text{otherwise.} \end{cases} \tag{1.28}$$

As an example of generating function, we consider the triangular function:

$$A_0(t) = \begin{cases} 0 & t < 1 \\ t + 1 & -1 \le t \le 0 \\ 1 - t & 0 \le t \le 1 \\ 0 & t > b \end{cases} \tag{1.29}$$

The basic functions of the generated h-uniform fuzzy partition are given by

$$A_k(x) = \begin{cases} 0 & x < x_k - h \\ \frac{x - x_k}{h} + 1 & x_k - h \le x \le x_k \\ 1 - \frac{x - x_k}{h} & x_k \le x \le x_k + h \\ 0 & x > x_k + h \end{cases} \quad k = 1, \ldots, n \tag{1.30}$$

As another example of generating function is the following:

$$A_0(t) = \frac{1}{2}[1 + \cos(\pi t)] \tag{1.31}$$

This function ranges between 0 (for $\cos(\pi t) = -1$) and 1 (for $\cos(\pi t) = 1$). The basic functions generated by (1.31) are given by

$$A_k(x) = \frac{1}{2}\left[1 + \cos\left(\frac{\pi}{h}(x - x_k)\right)\right] \quad k = 1, \ldots, n \qquad (1.32)$$

1.4 Direct and Inverse Fuzzy Transform

Let $\{A_1, A_2, \ldots, A_n\}$ be a fuzzy partition of $[a, b]$ and $f(x)$ be a continuous function on $[a, b]$. Thus we can consider the following real numbers for $i = 1, \ldots, n$:

$$F_k = \frac{\int_a^b f(x) A_k(x) dx}{\int_a^b A_k(x) dx} \quad k = 1, \ldots, n \qquad (1.33)$$

The n-tuple $\{F_1, F_2, \ldots, F_n\}$ is called the *fuzzy transform* of f with respect to $\{A_1, A_2, \ldots, A_n\}$. The F_k's are called *components* of the F-transform.

If the fuzzy partition $\{A_1, A_2, \ldots, A_n\}$ is uniform with nodes $\{x_1, x_2, \ldots, x_n\}$, the components are given (cf. [5], Lemma 1]) by the formula:

$$F_k = \begin{cases} \frac{2}{h} \int\limits_{x_1}^{x_2} f(x) A_k(x) dx & \text{if } k = 1 \\ \frac{1}{h} \int\limits_{x_{i-1}}^{x_i} f(x) A_k(x) dx & \text{if } k = 2, \ldots, n-1 \\ \frac{2}{h} \int\limits_{x_{n-1}}^{x_n} f(x) A_k(x) dx & \text{if } k = n \end{cases} \qquad (1.34)$$

Now we define the following function on $[a, b]$ given by a weighted average of the basic functions in which the weights are the F-transform components:

$$f_{F,n}(x) = \sum_{k=1}^{n} F_k A_k(x) \quad x \in [a, b] \qquad (1.35)$$

It is called *inverse F-transform* of f with respect to the uniform fuzzy partition $\{A_1, A_2, \ldots, A_n\}$. An important theorem [5] proves that the function $f_{F,n}$ approximates the continuous function f on $[a, b]$ with arbitrary precision. We enunciate below this theorem.

Theorem 1.1 Let $f(x)$ be a continuous function on $[a, b]$. For every $\varepsilon > 0$, there exists an integer $n(\varepsilon)$ and a related fuzzy partition $\{A_1, A_2, \ldots, A_{n(\varepsilon)}\}$ of $[a, b]$ such that for all $x \in [a, b]$:

$$\left| f(x) - f_{F,n(\varepsilon)}(x) \right| < \varepsilon, \tag{1.36}$$

1.5   Discrete Fuzzy Transform and Sufficient Density Concept

Theorem 1.1 concerns the approximation of a known continuous function f, but in many cases we only know that the function f assumes determined values in a set of m points $p_1,\ldots, p_m \in [a, b]$.

We assume that the set P of these nodes is *sufficiently dense with respect to the fixed fuzzy partition*, i.e. for each $k = 1,\ldots, n$ there exists an index $j \in \{1,\ldots, m\}$ such that $A_k(p_j) > 0$. Then we can define $\{F_1, F_2, \ldots, F_n\}$ as the *discrete F-transform* of f with respect to $\{A_1, A_2, \ldots, A_n\}$, where each F_k is given by

$$F_k = \frac{\sum_{j=1}^{m} f(p_j) A_k(p_j)}{\sum_{j=1}^{m} A_k(p_j)} \quad k = 1, \ldots, n \tag{1.37}$$

Similar to (1.32), we call the *discrete inverse F-transform* of f with respect to $\{A_1, A_2, \ldots, A_n\}$ to be the following function defined in the same points $p_1,\ldots, p_m$ of $[a, b]$:

$$f_{F,n}(p_j) = \sum_{k=1}^{n} F_k A_k(p_j) \tag{1.38}$$

Analogously to Theorem 1.1, we have the following approximation theorem [5], Theorem 5).

Theorem 1.2 Let $f(x)$ be a function assigned on a set P of points $p_1,\ldots, p_m$ of $[a, b]$. Then for every $\varepsilon > 0$, there exist an integer $n(\varepsilon)$ and a related fuzzy partition $\{A_1, A_2, \ldots, A_{n(\varepsilon)}\}$ of $[a, b]$ such that P is sufficiently dense with respect to $\{A_1, A_2, \ldots, A_{n(\varepsilon)}\}$ and for every $p_j \in [a, b], j = 1,\ldots, m$:

$$\left| f(x) - f_{F,n(\varepsilon)}(x) \right| < \varepsilon \tag{1.39}$$

Compliance with the constraint of sufficient density with respect to the partition is essential to ensure the existence of the discrete F-transform of f. In fact, if there exists a fuzzy set A_k of the fuzzy partition for which $A_k(p_j) = 0$ for every $j \in \{1,\ldots, m\}$, then (1.34) cannot be applied to calculate the F-transform component F_k. The meaning of this is that the fuzzy partition of the domain $[a, b]$ is too fine with respect to the dataset of the measures of the function f.

Now we show an example of a dataset of points not sufficiently dense with respect to the fuzzy partition. We consider an h-uniform fuzzy partition of the interval $[a, b]$

Table 1.1 The values of the 5 nodes in h-uniform fuzzy partition $(h = 0.75)$

Node	Value
x_1	1.00
x_2	1.75
x_3	2.50
x_4	3.25
x_5	4.00

$= [1, 4]$ given by $n = 5$ basic functions as (1.29). The values of the 5 nodes in the h-fuzzy partition $(h = 0.75)$ are given in Table 1.1.

In Fig. 1.9 we show the five basic functions: we put on abscissas the x-coordinates of the 10 points forming the dataset. This dataset is not sufficiently dense with respect to the fuzzy partition since the membership degree of any point to the fuzzy set A_2 is equal to zero.

Now we consider a coarser grained h-uniform fuzzy partition f given by four basic functions $(h = 1)$. Table 1.2 shows the values of the nodes.

As shown in Fig. 1.10, the dataset of points is sufficiently dense with respect to this new h-uniform fuzzy partition.

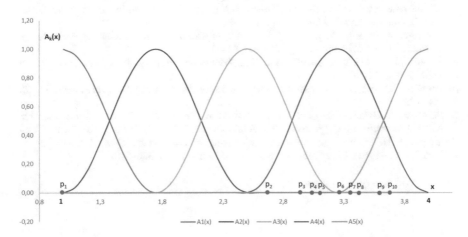

Fig. 1.9 Example of a dataset of points not sufficiently dense with respect to the fuzzy partition

Table 1.2 The values of the 4 nodes in h-uniform fuzzy partition $(h = 1)$

Node	Value
x_1	1.00
x_2	2.00
x_3	3.00
x_4	4.00

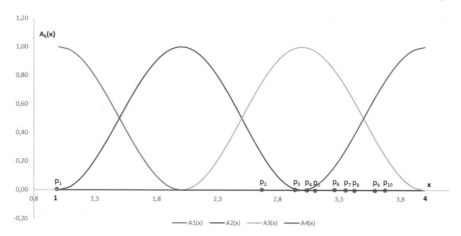

Fig. 1.10 Example of a dataset of points sufficiently dense with respect to the fuzzy partition

References

1. Adelson, E. H., Anderson, C. H., Bergen, J. R., Burt, J. R., & Ogden, J. (1984). Pyramid methods in image processing. *RCA Engineering, 29*(6), 33–41.
2. Chen, S. C., & Zhang, D. Q. (2004). Robust image segmentation using FCM with spatial constraints based on new kernel-induced distance measure. *IEEE Transactions on Systems, Man and Cybernetics, Part B, 34*(4), 1907–1916.
3. Dankova, M., & Valasek, R. (2006). Full fuzzy transform and the problem of image fusion. *Journal of Electrical Engineering, 12*(2006), 82–84.
4. Klement, E. P., Mesiar, R., & Pap E. (2000). *Triangular norms* (p. 387). Dordrecht: Kluwer. ISBN 0-7923-6416-3.
5. Perfilieva, I. (2006). Fuzzy transforms: theory and applications. *Fuzzy Sets and Systems, 157*(8), 993–1023.
6. Perfilieva, I., Haldeeva, E. (2001). Fuzzy transformation. In *IFSA World Congress and 20th NAFIPS International Conference, 2001*. Joint (9th edn, Vol. 4, pp. 1946–1948). IEEE.
7. Ruspini, E. H. (1969). A new approach to clustering. *Information and Control, 15,* 22–32.
8. Zadeh, L. A. (1965). Fuzzy sets. *Information and Control, 8*(3), 338–353.

Chapter 2
Multi-dimensional and High Degree Fuzzy Transform

The F-transform concept can be extended to approximate a function of more variables. Generally, we can approximate a continuous function f in n variables by using a multi-dimensional discrete inverse F-transform. In Sect. 2.1 we introduce the F-transform in two variables, used in many image analysis applications. The general multi-dimensional F-transform is introduced in Sect. 2.2.

2.1 Fuzzy Transform in Two Variables

Let a universe of discourse be composed from the rectangle $[a, b] \times [c, d]$ and let $n, m \geq 2, x_1, x_2, \ldots, x_n \in [a, b]$ and $y_1, y_2, \ldots, y_m \in [c, d]$ be $n + m$ assigned points, called *nodes*, such that $x_1 = a < x_2 < \cdots < x_n = b$ and $y_1 = c < \cdots < y_m = d$. Furthermore, let $A_1, \ldots, A_n: [a, b] \to [0, 1]$ be a uniform fuzzy partition of $[a, b]$, $B_1, \ldots, B_m: [a, b] \to [0, 1]$ be a uniform fuzzy partition of $[c, d]$ and $f(x, y): [a, b] \times [c, d] \to R$ be a continuous function on $[a, b] \times [c, d]$.

We define the $n \times m$ matrix $[F_{kl}]$ as the *F-transform* of the function f with respect to $\{A_1, \ldots, A_n\}$ and $\{B_1, \ldots, B_m\}$ given by the following components:

$$F_{kl} = \frac{\int_c^d \int_a^b f(x, y) A_k(x) B_l(y) \mathrm{d}x \mathrm{d}y}{\int_c^d \int_a^b A_k(x) B_l(y) \mathrm{d}x \mathrm{d}y} k = 1, \ldots, n \quad l = 1, \ldots, m \qquad (2.1)$$

Similar to the unidimensional case, we define the *inverse F-transform* of f with respect to $\{A_1, A_2, \ldots, A_n\}$ and $\{B_1, \ldots, B_m\}$ the following function on $[a, b] \times [c, d]$:

$$f_{nm}^F(x, y) = \sum_{k=1}^n \sum_{l=1}^m F_{kl} A_k(x) B_l(y) \qquad (2.2)$$

© Springer Nature Switzerland AG 2020
F. Di Martino and S. Sessa, *Fuzzy Transforms for Image Processing and Data Analysis*,
https://doi.org/10.1007/978-3-030-44613-0_2

A theorem, similar to Theorem 2.1 holds also in the case of two variables (cf [1], Theorem 14), so you can use (2.2) as an approximation of the function f.

Now we consider the bi-dimensional discrete case, assuming that the function f is known in a set of $N \times M$ points $(p_i, q_i) \in [a, b] \times [c, d]$, where $i = 1, ..., N$ and $j = 1, ..., M$.

This set of $N \times M$ points is sufficiently dense with respect to the fuzzy partition if, for each couple (k, l) with $k = 1, ..., n$ and $l = 1, ..., m$, there exists at least a point (p_i, q_i) such that $A_k(p_i) B_l(q_j) > 0$.

In this case we can define the bi-dimensional discrete F-transform of f with respect to $\{A_1, ..., A_n\}$ and $\{B_1, ..., B_m\}$, with components:

$$F_{kl} = \frac{\sum_{j=1}^{M} \sum_{i=1}^{N} f(p_i, q_j) A_k(p_i) B_l(q_j)}{\sum_{j=1}^{M} \sum_{i=1}^{N} A_k(p_i) B_l(q_j)} \quad k = 1, \ldots, n \quad l = 1, \ldots, m \quad (2.3)$$

We extend (2.2) to the case of two variables, by considering the *discrete inverse F-transform* of f with respect to $\{A_1, A_2, ..., A_n\}$ and $\{B_1, ..., B_m\}$ given by the following function defined in the same points $(p_j, q_j) \in [a, b] \times [c, d]$, with $i \in \{1, ..., N\}$ and $j \in \{1, ..., M\}$, as

$$f_{nm}^F(p_i, q_j) = \sum_{k=1}^{n} \sum_{l=1}^{m} F_{kl} A_k(p_i) B_l(q_j) \quad (2.4)$$

Analogous to the unidimensional case, one can prove that the discrete inverse F-transform (2.4) can approximate the original continuous function f with an arbitrary precision because the following theorem, generalization of Theorem 2.2 for the bi-dimensional case, holds:

Theorem 2.1 Let $f(x, y)$ be a function assigned on the set of points $(p_j, q_j) \in [a, b] \times [c, d]$, with $i = 1, ..., N$ and $j = 1, ..., M$. Then, for every $\varepsilon > 0$, there exist two integers $n(\varepsilon)$, $m(\varepsilon)$ and related fuzzy partitions $\{A_1, A_2, ..., A_{n(\varepsilon)}\}$ of $[a, b]$ and $\{B_1, B_2, ..., B_{m(\varepsilon)}\}$ of $[c, d]$ such that the set of points is sufficiently dense with respect to the fuzzy partitions $\{A_1, A_2, ..., A_{n(\varepsilon)}\}$ and $\{B_1, B_2, ..., B_{m(\varepsilon)}\}$ and we have for every $(p_j, q_j) \in [a, b] \times [c, d]$:

$$\left| f(p_i, q_j) - f_{n(\varepsilon)m(\varepsilon)}^F(p_i, q_j) \right| < \varepsilon \quad (2.5)$$

2.2 Multi-dimensional Fuzzy Transform

Now we extend the F-transform concept to approximate functions in s variables by considering a universe of discourse formed by the Cartesian product $[a_1, b_1] \times [a_2,$

$b_2] \times \cdots \times [a_s, b_s]$ of s given intervals $[a_i, b_i]$ $(i = 1, ..., s)$. Let $f: [a_1, b_1] \times [a_2, b_2] \times \cdots \times [a_s, b_s] \to R$ be a continuous function on the universe of discourse.

We define the $n_1 \times n_2 \times \cdots \times n_s$ matrix $[F_{k_1 k_s ... k_s}]$ as the *F-transform* of the function f with respect to $\{A_{11}, A_{12}, ..., A_{1n_1}\}, ..., \{A_{s1}, A_{s2}, ..., A_{sn_s}\}$ given by the following components:

$$F_{k_1 k_s ... k_s} = \frac{\int_{a_s}^{b_s} \cdots \int_{a_2}^{b_2} \int_{a_1}^{b_1} f(x_1, x_2, ..., x_s) A_{k_1}(x_1) A_{k_2}(x_2) ... A_{k_s}(x_s) dx_1 dx_2 ... dx_s}{\int_{a_s}^{b_s} \cdots \int_{a_2}^{b_2} \int_{a_1}^{b_1} A_{k_1}(x_1) A_{k_2}(x_2) ... A_{k_s}(x_s) dx_1 dx_2 ... dx_s}$$

(2.6)

with $k_1 = 1, ..., n_1, k_2 = 1, ..., n_2, \cdots, k_s = 1, ..., n_s$.

We define the *inverse F-transform* of f with respect to $\{A_{11}, A_{12}, ..., A_{1n_1}\}, ..., \{A_{s1}, A_{s2}, ..., A_{sn_s}\}$ the following function on $[a_1, b_1] \times [a_2, b_2] \times \cdots \times [a_s, b_s]$:

$$f_{n_1 n_2 ... n_s}^F(x_1, x_2, ..., x_s) = \sum_{k_1=1}^{n_1} \sum_{k_2=1}^{n_2} \cdots \sum_{k_s=1}^{n_s} F_{k_1 k_2 ... k_s} A_{k_1}(x_1) A_{k_2}(x_2) ... A_{k_s}(x_s)$$

(2.7)

Now we suppose that the function $f(x_1, x_2, ..., x_s)$ is known in N points $p_i = (p_{j1}, p_{j2}, ..., p_{js}) \in [a_1, b_1] \times [a_2, b_2] \times \cdots \times [a_s, b_s]$ for $j = 1, ..., s$.

We say that the set $P = \{(p_{11}, p_{12}, ..., p_{1s}), (p_{21}, p_{22}, ..., p_{2s}), \cdots, (p_{N1}, p_{N2}, ..., p_{Ns})\}$ is sufficiently dense with respect to the chosen partitions $\{A_{11}, A_{12}, ..., A_{1n_1}\}, ..., \{A_{s1}, A_{s2}, ..., A_{sn_s}\}$ if, for any combination $(h_1, ..., h_s) \in \{1, ..., n_1\} \times \cdots \times \{1, ..., n_s\}$, there is some $p_v = (p_{v1}, p_{v2}, p_{vs}) \in p$, $v \in \{1, ..., N\}$, such that $A_{1h_1}(p_{v1}) \cdot A_{2h_2}(p_{v2}) \cdot ... \cdot A_{sh_s}(p_{vs}) > 0$. So we can define the $(h_1, h_2, ..., h_s)$th component $F_{h_1 h_2 ... h_s}$ of the direct F-transform of f with respect to the basic functions $\{A_{11}, A_{12}, ..., A_{1n_1}\}, ..., \{A_{s1}, A_{s2}, ..., A_{sn_s}\}$ as

$$F_{h_1 h_2 ... h_s} = \frac{\sum_{j=1}^N f(p_{j1}, p_{j2}, ... p_{js}) \cdot A_{1h_1}(p_{j1}) \cdot A_{2h_2}(p_{j2}) \cdot ... \cdot A_{sh_s}(p_{js})}{\sum_{j=1}^N A_{1h_1}(p_{j1}) \cdot A_{2h_2}(p_{j2}) \cdot ... \cdot A_{sh_s}(p_{js})}$$

(2.8)

Now we define the *inverse multi-dimensional F-transform* of f with respect to the basic functions $\{A_{11}, A_{12}, ..., A_{1n_1}\}, ..., \{A_{s1}, A_{s2}, ..., A_{sn_s}\}$ to be the following function by setting for each point $p_j = (p_{j1}, p_{j2}, ..., p_{js}) \in [a_1, b_1] \times \cdots \times [a_s, b_s]$:

$$f_{n_1 n_2 ... n_s}^F(p_{j1}, p_{j2}, ..., p_{js}) = \sum_{h_1=1}^{n_1} \sum_{h_2=1}^{n_2} \cdots \sum_{h_s=1}^{n_s} F_{h_1 h_2 ... h_s} \cdot A_{1h_1}(p_{j1}) \cdot ... \cdot A_{sh_s}(p_{js})$$

(2.9)

for $j = 1, ..., N$. The following theorem, which is an extension of Theorem 2.2, holds:

Theorem 2.2 Let $f(x_1, \ldots, x_s)$ be a function assigned on the set of points $P = \{(p_{11}, p_{12}, \ldots, p_{1s}), (p_{21}, p_{22}, \ldots, p_{2s}), \ldots, (p_{m1}, p_{m2}, \ldots, p_{ms})\} \subset [a_1, b_1] \times [a_2, b_2] \times \cdots \times [a_k, b_k]$ and assuming values in $[0, 1]$. Then for every $\varepsilon > 0$, there exist k integers $n_1(\varepsilon), \ldots, n_s(\varepsilon)$ and related fuzzy partitions $\{A_{11}, A_{12}, \ldots, A_{1n_1(\varepsilon)}\}, \ldots, \{A_{s1}, A_{s2}, \ldots, A_{sn_s(\varepsilon)}\}$ such that the set P is sufficiently dense with respect to this fuzzy partitions. Moreover, for every $p_j = (p_{j1}, p_{j2}, \ldots, p_{js}) \in P, j = 1, \ldots, m$, the following inequality holds:

$$\left| f(p_{j1}, p_{j2}, \ldots, p_{js}) - f^F_{n_1(\varepsilon)n_2(\varepsilon)\ldots n_s(\varepsilon)}(p_{j1}, p_{j2}, \ldots, p_{js}) \right| < \varepsilon \qquad (2.10)$$

In other words, this theorem approximates the original continuous function f in a point with an arbitrary precision.

2.3 Sufficient Density in Multi-Dimensional Fuzzy Transforms

If the data are sufficiently dense with respect to the fuzzy partitions, we can apply the multi-dimensional inverse fuzzy transform (2.9) in order to approximate the function f in any point of the universe of discourse. The function s in a point of coordinates $(x_1, x_2, \ldots, x_s)$ can be approximate by

$$f^F_{n_1 n_2 \ldots n_s}(x_1, x_2, \ldots, x_s) = \sum_{h_1=1}^{n_1} \sum_{h_2=1}^{n_2} \cdots \sum_{h_s=1}^{n_s} F_{h_1 h_2 \ldots h_s} \cdot A_{1 h_1}(x_1) \cdot \ldots \cdot A_{s h_s}(x_s)$$

$$(2.11)$$

The sufficient density represents a strong constraint in Theorem 2.2. In order to check if the dataset of points is sufficiently dense with respect to the fuzzy partitions, we should verify that for any combination $(h_1, \ldots, h_s) \in \{1, \ldots, n_1\} \times \ldots \times \{1, \ldots, n_s\}$ exists at least a point p_v such that $A_{1h_1}(p_{v1}) \cdot A_{2h_2}(p_{v2}) \cdot \ldots \cdot A_{s h_s}(p_{vs}) > 0$.

Partition of $[a_1, b_1]$ is created by setting the n_1 nodes $x_{11} = a_1, x_{12}, \ldots, x_{1n1} = b_1$ and the uniform fuzzy partition of $[a_2, b_2]$ is created by setting the n_2 nodes $x_{21} = a_2, x_{22}, \ldots, x_{2n2} = b_2$. In Fig. 2.1 the nodes of the two fuzzy partitions are labeled on the two axes and the point of the dataset are plotted as red dots. As visible in Fig. 8, the data are not sufficiently dense with respect to the fuzzy partitions: indeed, in the gray area there are no points p_v such that $x_{1,i-1} < p_{v1} < x_{1,i+1}$ and $x_{2,j-1} < p_{v2} < x_{2,i+1}$. This result means that the uniform fuzzy partition is too fine with respect to the data.

To verify that the condition of sufficient density of the data with respect to the uniform fuzzy partitions is respected and to determine a suitable uniform fuzzy partitioning, it is necessary to implement an iterative process in which, starting with the most coarse-grained uniform fuzzy partitions of the universe, obtained considering

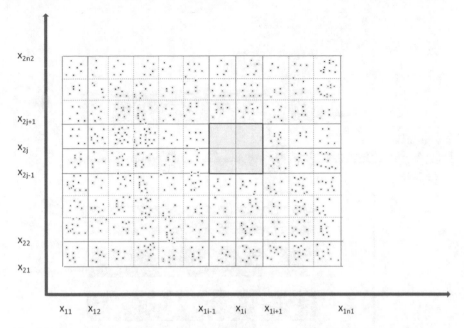

Fig. 2.1 Example of a dataset of points sufficiently dense with respect to the fuzzy partition

the minimum number of nodes $n_1 = n_2 = \cdots = n_s = 3$, we check if the data if sufficiently dense with respect to the fuzzy partitions and calculate the multi-dimensional inverse F-transform in any point: the difference between the inverse F-transform and the value of the function f in the point data gives the error of the approximation.

An example of approximation error is the Root Mean Square Error (RMSE) given by the well-known formula:

$$\text{RMSE} = \sqrt{\sum_{j=1}^{N} \left(f(p_{j1}, p_{j2}, \ldots, p_{js}) - f^{F}_{n_1 n_2(\varepsilon) \ldots n_s}(p_{j1}, p_{j2}, \ldots, p_{js}) \right)^2} \quad (2.12)$$

The iterative process stops if the approximation error is below a defined threshold: this means that the use of the multi-dimensional fuzzy transform to approximate the function f is acceptable. Otherwise, the process is iterated by considering fuzzy finer partitions. In Fig. 2.2 we schematize the flux diagram of this process.

Initially the number of nodes is set to 3 in any uniform fuzzy partition; moreover, the value of a parameter α is set as threshold of acceptability of the error of the approximation of the function f with the inverse F-transform, calculated in the data points. Then, the uniform fuzzy partitions are created and a check to verify if the dataset of points sufficiently dense with respect to the fuzzy partition is performed. If the data are not sufficiently dense with respect to the fuzzy partition, the process is stopped and it terminates with a message error indicating that the size of data is not

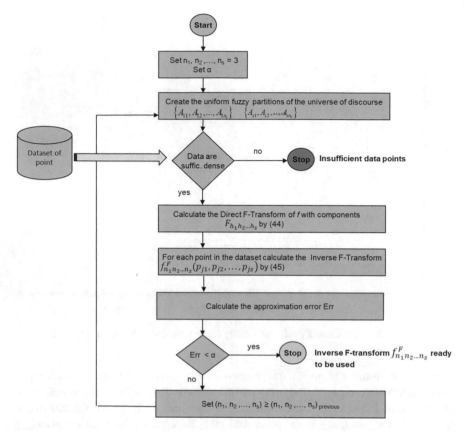

Fig. 2.2 Flux diagram of the iterative process implemented to check the best suitable inverse F-transform

sufficient to calculate a suitable approximation of the function f by the inverse F-transform. Else, the components of the direct F-transform and the inverse F-transform in any data point are calculated in order to evaluate the approximation error, if this error is less than the threshold value α, then the process stops ending with a message indicating that the calculated multi-dimensional inverse F-transform can be used to approximate the function f. Else, a finer set of fuzzy partitions is generated by setting the number of nodes in the interval of each variable greater or equal to the previous number of nodes.

2.4   High Degree Fuzzy Transform

In this section, we introduce the concept of fuzzy transform of higher degree or F^r-transform. Let $A_k(x)$ be a basic function of $[a, b]$, $k = 1, \ldots, n$, and $L_2([x_{k-1}, x_{k+1}])$

be the Hilbert space of square integrable functions $g: [x_{k-1}, x_{k+1}] \to R$ (reals) with weighted inner product:

$$\langle f, g \rangle_k = \frac{\int_{x_{k-1}}^{x_{k+1}} f(x)g(x)A_k(x)dx}{\int_{x_{k-1}}^{x_{k+1}} A_k(x)dx} \tag{2.13}$$

where $f, g \in L_2([x_{k-1}, x_{k+1}]$ and A_k is the weight function.

Let $L_2^r([x_{k-1}, x_{k+1}])$, with r positive integers, be a linear subspace of the Hilbert spaces $L_2([x_{k-1}, x_{k+1}])$ with orthogonal basis given by polynomials $P_k^0, P_k^1, \ldots, P_k^r$ obtained applying the Gram–Schmidt orthonormalization [2] to the linear independent system of polynomials $\{1, x, x^2, \ldots, x^r\}$ defined in the interval $[x_{k-1}, x_{k+1}]$. We have that

$$\begin{cases} P_k^0 = 1 \\ P_k^{i+1} = x^{i+1} - \sum_{j=1}^{i} \frac{\langle x^{i+1}, P_k^j \rangle}{\langle P_k^j, P_k^j \rangle} & i = 1, \ldots, r-1 \end{cases} \tag{2.14}$$

The following Lemma holds (cf, [3], Lemma 2.3):

Lemma 2.3 Let F_k^r be the orthogonal projection of the function f on $L_2^r([x_{k-1}, x_{k+1}])$. Then for $k = 1, \ldots, n$:

$$F_k^r(x) = \sum_{i=1}^{r} c_{k,i} P_k^i(x) \tag{2.15}$$

where

$$c_{k,i} = \frac{\langle f, P_k^i \rangle_k}{\langle P_k^i, P_k^i \rangle_k} = \frac{\int_{x_{k-1}}^{x_{k+1}} f(x) P_k^i(x) A_k(x)dx}{\int_{x_{k-1}}^{x_{k+1}} \left(P_k^i(x)\right)^2 A_k(x)dx} \tag{2.16}$$

F_k^r is called the kth component of *the direct F^r-transform component* of f and we put $F^r[f] = \left(F_1^r, F_2^r, \ldots, F_n^r\right)$.

For $r = 0$, we have $P_k^0 = 1$ and the F^0-*transform* is given by the F-transform in one variable $(F_k^0(x) = c_{k,0})$.

We can define the *inverse F^r-transform* of f in a point x as

$$f_{F,n}^r(x) = \sum_{k=1}^{n} F_k^r A_k(x) \quad x \in [a, b] \tag{2.17}$$

Likewise, we define the Hilbert space $L_2([x_{k-1}, x_{k+1}] \times [y_{l-1}, y_{l+1}])$ of square integrable two-variable functions $f: [x_{k-1}, x_{k+1}] \times [y_{l-1}, y_{l+1}] \to R$ with weighted inner product:

$$\langle f, g \rangle_{kl} = \int\limits_{x_{k-1}}^{x_{k+1}} \int\limits_{y_{l-1}}^{y_{l+1}} f(x, y) g(x, y) A_k(x) A_l(y) dy dx \qquad (2.18)$$

Two functions $f, g \in L_2([x_{k-1}, x_{k+1}] \times [y_{l-1}, y_{l+1}])$ are orthogonal if $\langle f, g \rangle_{kl} = 0$.

Let $L_2^p([x_{k-1}, x_{k+1}])$ and $L_2^r([y_{l-1}, y_{l+1}])$, $p, r \geq 0$ be two linear subspaces of $L_2([x_{k-1}, x_{k+1}])$ and $L_2([y_{l-1}, y_{l+1}])$ with orthogonal basis given by polynomials $\{P_k^i(x)\}_{i=0,\ldots,p}$ and $\{Q_l^j(y)\}_{j=0,\ldots,r}$, constructed via (2.14), respectively.

Let $s \geq 0$ be a positive integer and let's consider all pairs (i, j) such that $0 \leq i + j \leq s$. We introduce a linear subspace $L_2^s(([x_{k-1}, x_{k+1}] \times [y_{l-1}, y_{l+1}])$ of $L_2([x_{k-1}, x_{k+1}] \times [y_{l-1}, y_{l+1}])$ having the following orthogonal basis:

$$\left\{ S_{kl}^{ij}(x, y) = P_k^i(x) Q_l^j(y) \right\}_{i=0\ldots,p;\, j=0,\ldots,r:\, i+j \leq s} \qquad (2.19)$$

where s is the maximum degree of polynomials $P_k^i(x) Q_l^j(y)$.

The following Lemma holds (cf, [4], Lemma 2.5):

Lemma 2.4 Let $f(x, y) \in L_2([x_{k-1}, x_{k+1}] \times [y_{l-1}, y_{l+1}])$. Then the orthogonal projection of f on $L_2^s(([x_{k-1}, x_{k+1}] \times [y_{l-1}, y_{l+1}])$, $s \geq 0$, is the polynomial of degree s given by

$$F_{kl}^s(x, y) = \sum_{0 \leq i+j \leq s} c_{kl}^{ij} S_{kl}^{ij}(x, y) \qquad (2.19)$$

where the coefficients c_{kl}^{ij} are given by

$$c_{kl}^{ij} = \frac{\langle f, S_{kl}^{ij} \rangle_{kl}}{\langle S_{kl}^{ij}, S_{kl}^{ij} \rangle_{kl}} = \frac{\int_{y_{l-1}}^{y_{l+1}} \int_{x_{k-1}}^{x_{k+1}} f(x, y) S_{kl}^{ij}(x, y) A_k(x) B_l(y) dx dy}{\int_{y_{l-1}}^{y_{l+1}} \int_{x_{k-1}}^{x_{k+1}} \left(S_{kl}^{ij}(x, y) \right)^2 A_k(x) B_l(y) dx dy} \qquad (2.20)$$

For $s = 0$, we have $S_{kl}^{00}(x, y) = 1$ and (2.19) correspond to the F-transform component

$$F_{kl}^0(x, y) = c_{kl}^{00} S_{kl}^{00}(x, y) = \frac{\int_{y_{l-1}}^{y_{l+1}} \int_{x_{k-1}}^{x_{k+1}} f(x, y) A_k(x) B_l(y) dx dy}{\int_{y_{l-1}}^{y_{l+1}} \int_{x_{k-1}}^{x_{k+1}} A_k(x) B_l(y) dx dy} \qquad (2.21)$$

2.5 F-Fuzzy Transform

In this section we introduce the concept of F^1 transform obtained by considering an orthogonal basis given by polynomials with maximum degree $s = 1$. The concept of F^1-transform can be used in image analysis to improve the performance of the F-transform method.

Let A_k, $k = 1, ..., n$, be an h-uniform generalized fuzzy partition of $[a, b]$ and $f(x, y) \in L_2([a, b])$. We consider the linear subspace $L_2^1([a, b])$ of $L_2([a, b])$ with orthogonal basis given by the polynomial:

$$S_k^0(x) = P_k^0(x) = 1$$
$$S_k^1(x) = P_k^1(x) = x - x_k \tag{2.22}$$

Then the kth element of the F^1-transform is given by

$$F_k^1(x) = c_k^0 S_k^0(x) + c_k^1 S_k^1(x) = c_k^0 + c_k^1(x - x_k) \tag{2.23}$$

where the three coefficients c_k^0 and c_k^1 are given by

$$c_k^0 = \frac{\langle f, S_k^0 \rangle_k}{\langle S_k^0, S_{kl}^0 \rangle_k} = \frac{\int_{x_{k-1}}^{x_{k+1}} f(x) A_k(x) dx}{\int_{x_{k-1}}^{x_{k+1}} A_k(x) dx} \tag{2.24}$$

and

$$c_k^1 = \frac{\langle f, S_k^1 \rangle_k}{\langle S_k^1, S_k^1 \rangle_k} = \frac{\int_{x_{k-1}}^{x_{k+1}} f(x) A_k(x) dx}{\int_{x_{k-1}}^{x_{k+1}} A_k(x)(x - x_k)^2 dx} \tag{2.25}$$

The following theorem holds [3], Theorem 2.1.

Theorem 2.5 Let $f(x) \in L_2([a, b])$ and $\{A_k(x), k = 1, ..., n\}$ be a generalized fuzzy partition of $[a, b]$. Moreover let f and A_k be four times continuously differentiable on $[a, b]$ Then, for each $k = 1, ..., n$, the following approximation holds true:

$$c_k^1 = f_k' + O(h) \tag{2.26}$$

From Theorem 2.5 the following corollary [3], Corollary 2.6, follows.

Corollary 2.6 Let $f(x) \in L_2([a, b])$ and $\{A_k(x), k = 1, ..., n\}$ be a generalized fuzzy partition of $[a, b]$. Moreover let f and A_k be four times continuously differentiable on $[a, b]$ Then, for each $k = 1, ..., n$, holds:

$$f(x) = F_k^1 + O(h^2) \quad x_{k-1} \le x \le x_{k+1} \tag{2.27}$$

Now consider a bi-dimensional function $f(x, y) \in L_2([a, b] \times [c, d])$.

Let $\{(A_k, B_l), k = 1, \ldots, n, l = 1, \ldots, m\}$ be an h-uniform generalized fuzzy partition of $[a, b] \times [c, d]$. For $s = 1$, the orthogonal basis of the linear subspace $L_2^1([a, b] \times [c, d])$ is obtained with the following polynomials:

$$S_{kl}^{00}(x, y) = P_k^0(x)Q_l^0(y) = 1$$
$$S_{kl}^{10}(x, y) = P_k^1(x)Q_l^0(y) = x - x_k$$
$$S_{kl}^{01}(x, y) = P_k^0(x)Q_l^1(y) = y - y_l \qquad (2.28)$$

The orthogonal projection of f on $L_2^1([a, b] \times [c, d])$ is given by

$$F_{kl}^1(x, y) = \sum_{0 \le i+j \le 1} c_{kl}^{ij} S_{kl}^{ij}(x, y) = c_{kl}^{00} + c_{kl}^{10}(x - x_k) + c_{kl}^{01}(y - y_l) \qquad (2.29)$$

where the three coefficients $c_{kl}^{00}, c_{kl}^{10}, c_{kl}^{01}$ are given by (7) and Lemma 2.3:

$$c_{kl}^{00} = \frac{\langle f, S_{kl}^{00}\rangle_{kl}}{\langle S_{kl}^{00}, S_{kl}^{00}\rangle_{kl}} = \frac{\int_{y_{l-1}}^{y_{l+1}} \int_{x_{k-1}}^{x_{k+1}} f(x, y) A_k(x) B_l(y) dx dy}{\int_{x_{k-1}}^{x_{k+1}} A_k(x) dx \int_{y_{l-1}}^{y_{l+1}} B_l(y) dy} \qquad (2.30)$$

$$c_{kl}^{10} = \frac{\langle f, S_{kl}^{10}\rangle_{kl}}{\langle S_{kl}^{10}, S_{kl}^{10}\rangle_{kl}} = \frac{\int_{y_{l-1}}^{y_{l+1}} \int_{x_{k-1}}^{x_{k+1}} f(x, y)(x - x_k) A_k(x) B_l(y) dx dy}{\int_{x_{k-1}}^{x_{k+1}} A_k(x)(x - x_k)^2 dx \int_{y_{l-1}}^{y_{l+1}} B_l(y) dy} \qquad (2.31)$$

$$c_{kl}^{01} = \frac{\langle f, S_{kl}^{01}\rangle_{kl}}{\langle S_{kl}^{01}, S_{kl}^{01}\rangle_{kl}} = \frac{\int_{y_{l-1}}^{y_{l+1}} \int_{x_{k-1}}^{x_{k+1}} f(x, y)(y - y_l) A_k(x) B_l(y) dx dy}{\int_{x_{k-1}}^{x_{k+1}} A_k(x) dx \int_{y_{l-1}}^{y_{l+1}} B_l(y)(y - y_l)^2 dy} \qquad (2.32)$$

The coefficients of the F^1-transform can be used to approximate the function $f(x, y)$ up to the first order. The following lemma holds [4], Lemma 2.9:

Lemma 2.7 Let $f(x, y) \in L_2([a, b] \times [c, d])$ and $\{A_k, B_l\}, k = 1, \ldots, n, l = 1, \ldots, m\}$, $n, m \ge 2$, be an h-generalized fuzzy partition of $[a, b] \times [c, d]$. Moreover let f be twice continuously differentiable on $[a, b] \times [c, d]$ and A_k (resp., B_l) be twice continuously differentiable on $[a, b]$(resp., $[c, d]$). Then the following holds:

$$c_{kl}^{00} = f(x_k, y_l) + O(h^2) \quad k = 1, \ldots, n \quad l = 1, \ldots, m \qquad (2.34)$$

The following theorem holds [4, Theorem 2.10]:

Theorem 2.8 Let $f(x, y) \in L_2([a, b] \times [c, d])$ and $\{A_k(x), B_l(y) \ k = 1, \ldots, n, l = 1, \ldots, m\}$, $n, m \ge 2$ be a generalized fuzzy partition of $[a, b] \times [c, d]$. Moreover, let f be four times continuously differentiable on $[a, b] \times [c, d]$ and A_k (resp., B_l) four times continuously differentiable $[a, b]$ (resp., $[c, d]$).

Then, for each $k = 1, \ldots, n, l = 1, \ldots, n$, there exist $x_{kl} \in (x_k - h, x_k + h)$ and $y_{kl} \in (y_l - h, y_l + h)$ such that

$$c_{kl}^{10} = \frac{\partial f}{\partial x}(x_k, y_{kl}) + O(h^2)$$

$$c_{kl}^{01} = \frac{\partial f}{\partial y}(x_{kl}, y_l) + O(h^2) \qquad (2.35)$$

Let $\left(F_{11}^1, \ldots, F_{nm}^1\right)$ be the F^1-trasform of f with respect to an h-generalized fuzzy partition. Then the following holds [4]. Corollary 11):

$$f(x, y) = c_{kl}^{00} + c_{kl}^{10}(x - x_k) + c_{kl}^{01}(y - y_l) + O(h^2) = F_{kl}^1(x, y) + O(h^2) \qquad (2.36)$$

The bi-dimensional inverse F^1-transform of the function f is given by $\hat{f}_{nm}^1$: $[a, b] \times [c, d] \rightarrow R$, where

$$\hat{f}_{nm}^1(x, y) = \frac{\sum_{k=1}^n \sum_{l=1}^m F_{nm}^1(x, y) A_k(x) B_l(y)}{\sum_{k=1}^n \sum_{l=1}^m A_k(x) B_l(y)} \qquad (2.37)$$

The following theorem allows to use the inverse F^1-transform $\hat{f}_{nm}^1$ to estimate the function f [4, Theorem 2.14]:

Theorem 2.9 Let $\{A_k(x) \times B_l(y) \, k = 1, \ldots, n, l = 1, \ldots, m\}$ $n, m \geq 2$ be an h-uniform generalized fuzzy partition of $[a, b] \times [c, d]$ and be $\hat{f}_{nm}^1$ the inverse F^1-transform of f with respect to this fuzzy partition. Moreover, be f, A_k, B_l four times continuously differentiable on $[a, b] \times [c, d]$. Then, for each couple $(x, y) \in [a, b] \times [c, d]$ holds:

$$f(x_k, y_l) - \hat{f}_{nm}^1(x, y) = O(h^2) \qquad (2.38)$$

This theorem allows us to approximate a two-dimensional function four times continuously differentiable on the interval $[a, b] \times [c, d]$ with the inverse F^1-transform up to $O(h^2)$.

References

1. Perfilieva, I. (2006). Fuzzy transforms: theory and applications. *Fuzzy Sets and Systems, 157*(8), 993–1023.
2. Hazewinkel, M. (Eds.) (2000). *Encyclopaedia of Mathematics, Supplement* (Vol. II, p. 526). Netherlands: Springer. https://doi.org/10.1007/978-94-015-1279-4.
3. Perfilieva, I., Daňková, M., & Bede, B. (2011). Towards a higher degree F-transform. *Fuzzy Sets and Systems, 180*, 3–19.
4. Perfilieva, I., Hodáková, P., & Hurtik, P. (2016). Differentiation by the F-transform and application for edge detection. *Fuzzy Sets and Systems, 288*, 96–114.

Chapter 3
Fuzzy Transform for Image and Video Compression

In this chapter, methods based on F-transform are explored and applied to image and video compression. They are considered lossy compression methods, in which the image is rebuilt with a loss of information. An example of a famous lossy image compression algorithm is the Joint Photographic Experts Group (JPEG) method.

Let I be a gray image with L gray levels given by a matrix with $N \times M$ cells called *pixels*. The value of any pixel is an integer between 0 and $L - 1$. The direct F-transform is used to compress the image, then the decoded image is approximated with the inverse F-transform. This method is lossy compression.

3.1 Coding and Decoding Images by Using Bi-Dimensional F-Transforms

Let $R: (i, j) \in \{1, ..., N\} \times \{1, ..., M\} \to [0, 1]$, $R(i, j)$ being the normalized value of the pixel $I(i, j)$, that is $R(i, j) = I(i, j)/L$.

Let $\{A_1, A_2, ..., A_n\}$ be a uniform fuzzy partition of $[1, N]$ and $\{B_1, B_2, ..., B_m\}$ be a uniform fuzzy partition of $[1, M]$, with $n < N$ and $m < M$.

In [14] the image R is compressed by using a bi-dimensional discrete direct F-transform in two variables $[F_{kl}]$ defined for each $k = 1, ..., n$ and $l = 1, ..., m$, as

$$F_{kl} = \frac{\sum_{j=1}^{M} \sum_{i=1}^{N} R(i, j) A_k(i) B_l(j)}{\sum_{j=1}^{M} \sum_{i=1}^{N} A_k(i) B_l(j)} \tag{3.1}$$

It is meaningful to note that in this case, the data are always sufficiently dense with respect to the fuzzy partitions because consecutive data points are equidistant from each other and their distance is 1. The distance between two consecutive nodes is $h = (b - a)/(n + 1) = (N - 1)/(n + 1) > 1$ because $n < N$, then for any $k = 1, ..., N$, there is at least a value i, $i = 1, ..., N$, such that $A_k(i) > 0$. For the same reason,

© Springer Nature Switzerland AG 2020
F. Di Martino and S. Sessa, *Fuzzy Transforms for Image Processing and Data Analysis*,
https://doi.org/10.1007/978-3-030-44613-0_3

there is at least a value $j, j = 1, ..., M$, such that $B_l(j) > 0$. Thus, the denominator in (3.1) is never null.

The compressed image (3.1) can be decoded by using the bi-dimensional inverse discrete F-transform for every $(i, j) \in \{1, ..., N\} \times \{1, ..., M\}$.

$$R^F_{nm}(i, j) = \sum_{k=1}^{n} \sum_{l=1}^{m} F_{kl} A_k(i) B_l(j) \tag{3.2}$$

Equation (3.2) approximates the normalized image $R(i, j)$. The decoded image is given by

$$I^F_{nm}(i, j) = R^F_{nm}(i, j) \cdot L \tag{3.3}$$

In [3, 14] the cosinusoidal basic functions used are defined as

$$A_1(x) = \begin{cases} 0.5\left(1 + \cos\frac{\pi}{h}(x - x_1)\right) & \text{if } x \in [x_1, x_2] \\ 0 & \text{otherwise} \end{cases}$$

$$A_k(x) = \begin{cases} 0.5\left(1 + \cos\frac{\pi}{h}(x - x_k)\right) & \text{if } x \in [x_{k-1}, x_{k+1}] \\ 0 & \text{otherwise} \end{cases}$$

$$A_n(x) = \begin{cases} 0.5\left(1 + \cos\frac{\pi}{h}(x - x_n)\right) & \text{if } x \in [x_{n-1}, x_n] \\ 0 & \text{otherwise} \end{cases} \tag{3.4}$$

where $k = 2, ..., n - 1$, $h = (N - 1)/(n - 1)$,

$$B_1(y) = \begin{cases} 0.5\left(1 + \cos\frac{\pi}{s}(y - y_1)\right) & \text{if } y \in [y_1, y_2] \\ 0 & \text{otherwise} \end{cases}$$

$$B_t(y) = \begin{cases} 0.5\left(1 + \cos\frac{\pi}{s}(y - y_t)\right) & \text{if } y \in [y_{t-1}, y_{t+1}] \\ 0 & \text{otherwise} \end{cases}$$

$$B_m(y) = \begin{cases} 0.5\left(1 + \cos\frac{\pi}{s}(y - y_m)\right) & \text{if } y \in [y_{m-1}, y_m] \\ 0 & \text{otherwise} \end{cases} \tag{3.5}$$

where $t = 2, ..., m - 1$, $s = (M - 1)/(m - 1)$. Below, in the pseudocode, we show the coding and decoding images via F-transform.

Algorithm: F-transform *Image Coding*	
Input:	*Source Image I*
Output:	*Direct F-transform components*
1	Set n,m
2	Normalize the source image I obtaining the image R
3	Set the basic functions A_1, A_2,... A_n and B_1, B_2,... B_n
4	Calculate the direct F-transform via (3.1)
5	Store the direct F-transform components

Algorithm: F-transform *Image Decoding*	
Input:	*Direct F-transform*
Output:	*Decoded image*
1	For each i = 1 to N
2	For each j = 1 to M
3	Calculate the inverse F-transform value $R^F_{nm}(i,j)$ via (3.2)
4	Calculate the decoded pixel $I^F_{nm}(i,j)$ via (3.3)
5	Next j
6	Next i
7	Store the decoded image

To measure the compression level of the image, we define the compression rate $\rho = nm/NM$ given by the inverse of the compression ratio *NM/nm*. The greater the compression of the image, that is, the smaller the compression rate, the more degraded the decompressed image will be compared to the original one.

An index, widely used to measure the quality of the reconstructed image, is the Peak Signal to Noise Ratio (shortly, PSNR) given by

$$PSNR = 20 \log_{10} \frac{L}{RMSE} \tag{3.6}$$

where RMSE is the Root Mean Square Error, given by the formula:

$$RMSE = \sqrt{\frac{\sum_{i=1}^{N} \sum_{j=1}^{M} (R(i, j) - R^F_{NM}(i, j))^2}{N \times M}} \tag{3.7}$$

The PSNR index can be applied to verify if the quality of the reconstructed image is acceptable.

3.2 Image Compression with Block Decompositions

In [3] the normalized image is partitioned into submatrices R_B of sizes $N(B) \times M(B)$, called *blocks* [1, 2], where $N(B)$ is a positive integer less than N and divisible by N and $M(B)$ a positive integer less than M and divisible by M. Each block is compressed to a block F_B of sizes $n(B) \times m(B)$, where $n(B) < N(B)$ and $m(B) < M(B)$, via the discrete *F*-transform $\mathbf{F}_{n(B)m(B)}[R_B] = (F^B_{kl})$ whose components, for each $k = 1, ..., n(B)$ and $l = 1, ..., m(B)$, are given by

$$F^B_{kl} = \frac{\sum_{j=1}^{M(B)} \sum_{i=1}^{N(B)} R_B(i, j) A_k(i) B_l(j)}{\sum_{j=1}^{M(B)} \sum_{i=1}^{N(B)} A_k(i) B_l(j)} \tag{3.8}$$

The compressed block F_B is decoded to a block $R^F_{n(B)m(B)}$ of sizes $N(B) \times M(B)$ by using the bi-dimensional inverse discrete F-transform (3.2) for every $(i, j) \in \{1, ..., N_B\} \times \{1, ..., M_B\}$:

$$R^F_{n(B)m(B)}(i, j) = \sum_{k=1}^{n(B)} \sum_{l=1}^{m(B)} F^B_{kl} A_k(i) B_l(j) \tag{3.9}$$

Finally, the normalized decoded image is obtained merging the decoded blocks.

In [3] this block decomposition image compression method is applied, using the cosinusoidal basic functions (3.6) and (3.7) where N, M, $N(B)$, and $M(B)$ are substituted by $N(B)$, $M(B)$, $n(B)$, and $M(B)$, respectively. Below are shown the F-transform Block image coding and decoding algorithms.

Algorithm: F-transform *Block Image Coding*	
Input:	*Source Image I*
Output:	*Compressed normalized image*
1	Set N(B) M(B), n(B),m(B)
2	Normalize the source image I obtaining the image R
3	Divide the image in blocks with size N(B)× M(B)
4	For each block
5	Calculate the direct F-transform via (3.8)
6	Next
7	Join the compressed blocks to obtain the compressed normalized image
8	Store the compressed normalized image

Algorithm: F-transform *Block Image Decoding*	
Input:	*Compressed normalized image*
Output:	*Decoded image*
1	Divide the normalized image in blocks with size n(B)× m(B)
2	FOR each i = 1 to N(B)
3	FOR each j = 1 to M(B)
4	Calculate the inverse F-transform value $R^F_{n(B)m(B)}(i,j)$ via (3.9)
5	Calculate the decoded pixel via (3.3)
6	NEXT j
7	NEXT i
8	Join the decompressed blocks to obtain the decompressed image
9	Store the decoded image

In [3] comparison tests are performed between the F-transform image compression and the Fuzzy Relation Equation, JPEG and Discrete Cosine Transform (DCT) algorithm. The authors measure the PSNR index obtained by applying various compression rates for a sample of about a hundred gray images, showing that the quality of the images decoded via F-transform method is better than the ones obtained by Fuzzy Relation Equation and DCT algorithms and comparable with the ones obtained via the JPEG algorithm.

In Figs. 3.1, 3.2, 3.3, and 3.4 we show the 256×256 gray level original image "Bridge" and the decoded images obtained by using the Fuzzy Relation Equation, F-transform, and JPEG algorithms with a compression rate $\rho = 0.25$, respectively. The F-transform decoded image was obtained by using the block decomposition method, compressing any block 4×4 in a block 2×2.

Table 3.1 shows the PSNR values obtained for various compression rates applying the F-transform, Fuzzy Relation Equation, Discrete Cosine Transform, and JPEG

Fig. 3.1 Bridge original image

Fig. 3.2 Fuzzy Relation Equation

Fig. 3.3 *F*-transform

Fig. 3.4 DCT–JPEG

algorithms, respectively. The *F*-transform method gives better results with respect to Fuzzy Relation Equation and Discrete Cosine Transform algorithm and close to the ones obtained via the JPEG algorithm for strong compressions.

3.3 High Degree Fuzzy Transforms for Coding/Decoding Images

The high degree fuzzy transform can be applied to coding/decoding images. In [7] the discrete F^1-transform is used to compress gray images. Let $R: (i,j) \in \{1, ..., N\} \times \{1, ..., M\} \to [0, 1]$, $R(i,j)$ being the normalized value of the pixel $I(i,j)$, that is $R(i,$

Fig. 3.5 Lena original image

Table 3.1 PSNR values obtained for the image Bridge with different lossy compression methods

$\rho(B)$	PSNR in FTR	PSNR in FEQ	PSNR in DCT	PSNR in JPEG
0.03	20.7262	11.0283	18.6115	22.6985
0.06	21.4833	14.2812	19.4849	24.7253
0.14	23.2101	16.4632	20.8430	28.1149
0.25	24.6975	19.7759	22.5470	31.2148
0.44	27.0960	23.7349	26.1490	37.2367

$j) = I(i, j)/L$. The image R is compressed in an image of size $n \times m$ by considering the F^1-transform components:

$$F_{kl}^1 = c_{kl}^{00} + c_{kl}^{10}(i - k) + c_{kl}^{01}(j - l) \tag{3.10}$$

where

$$c_{kl}^{00} = \frac{\sum_{j=1}^{M} \sum_{i=1}^{N} R(i, j) A_k(i) B_l(j)}{\sum_{j=1}^{M} \sum_{i=1}^{N} A_k(i) B_l(j)} \tag{3.11}$$

$$c_{kl}^{10} = \frac{\sum_{j=1}^{M} \sum_{i=1}^{N} R(i, j)(i - k) A_k(i) B_l(j)}{\sum_{i=1}^{N} A_k(i)(i - k)^2 \sum_{j=1}^{M} B_l(j)} \tag{3.12}$$

$$c_{kl}^{01} = \frac{\sum_{j=1}^{M} \sum_{i=1}^{N} R(i, j)(j - l) A_k(i) B_l(j)}{\sum_{i=1}^{N} A_k(i) \sum_{j=1}^{N} B_l(j)(j - l)^2} \tag{3.13}$$

where $k = 1, \ldots, n, 1 = 1, \ldots, m$. The decoded image is given by the bi-dimensional inverse F^1-transform:

$$R_{nm}^{F^1}(i, j) = \hat{f}_{nm}^1(i, j) = \frac{\sum_{k=1}^{n} \sum_{l=1}^{m} F_{nm}^1(i, j) A_k(i) B_l(j)}{\sum_{k=1}^{n} \sum_{l=1}^{m} A_k(i) B_l(j)} \qquad (3.14)$$

Likewise, we can partition the original image in blocks, using the formulas (3.10) and (3.14) for coding and decoding each image block. In [7] the F^1-transform is applied to image blocks. In Figs. 3.6, 3.7, and 3.8, we show the decoded image obtained applying different compression rates for the gray image Lena (Fig. 3.5).

Comparison results between the quality of the decoded images obtained for various compression rates applying the F-transform and the F^1-transform method show

Fig. 3.6 F^1-tr decoded image ($\rho = 0.16$)

Fig. 3.7 F^1-tr decoded image ($\rho = 0.28$)

Fig. 3.8 F^1-tr decoded
image ($\rho = 0.44$)

that the difference of the PSNR index is greater than 0.1 only for not strong compression ($\rho \geq 0.25$). Conversely, for strong compression rates this difference is not
significant as the correction introduced by F^1-transform considering the first order
is made evanescent by the presence of noise in the compressed image. Figure 3.9
compares the trend of the PSNR index obtained for the gray image Lena applying
the F-transform and the F^1-transform method.

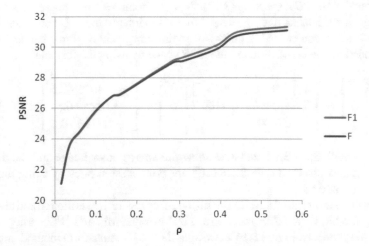

Fig. 3.9   PSNR obtained applying the F- and F^1-transform on the image Lena

3.4   Color Image Compression in the YUV Space

A color image is given by three gray level matrices defined in the *RGB* (red, green, and blue) space, where *RGB* refers to a system for representing the colors to be used on a computer display. A useful approach used in image compression is to convert the *RGB* space in the *YUV* color space given by the brightness component (the *Y* band) and by the two chrominance components (the *U* and *V* bands).

The advantage of using the *YUV* space in image compression is that the human eye is very sensitive to electromagnetic radiation in the brightness band and is not very sensitive to those in the two chrominance bands. Thus, in lossy color image compression, using the *YUV* space, it is possible to compress the image in the *U* and *V* bands more than in the *Y* band, without high loss of information. The standard JPEG image compression method [13] manages color images in the *YUV* space.

We are supposed to have a color image in the *RGB* space. The transformation matrix to convert the image in the *YUV* space is given by

$$
\begin{bmatrix} Y \\ U \\ V \end{bmatrix} = \begin{bmatrix} 0.299 & 0.587 & 0.114 \\ -0.169 & -0.332 & 0.500 \\ 0.500 & -0.419 & -0.813 \end{bmatrix} \begin{bmatrix} R \\ G \\ B \end{bmatrix} + \begin{bmatrix} 0 \\ 0.5 \\ 0.5 \end{bmatrix} \tag{3.15}
$$

Following the idea developed in [4–6, 11, 12], these authors propose a method based on *F*-transform for compressing color images in the *YUV* space. The image is transformed in the *YUV* space by (3.13), then two compression rates are used: ρ_Y for the image in the *Y* band, and $\rho_U = \rho_V < \rho_Y$ for the images in the *U* and *V* bands The discrete *F*-transform is used in coding/decoding images in the three bands. Finally, the decoded image is converted in the *RGB* space by using the formula:

$$
\begin{bmatrix} R \\ G \\ B \end{bmatrix} = \begin{bmatrix} 1 & 0 & 1.4075 \\ 1 & -0.3455 & -0.7169 \\ 1 & 1.7790 & 0 \end{bmatrix} \begin{bmatrix} Y \\ U \\ V \end{bmatrix} + \begin{bmatrix} -1.4075 \cdot 0.5 \\ 1.0624 \cdot 0.5 \\ -1.7790 \cdot 0.5 \end{bmatrix} \tag{3.16}
$$

Figures 3.10, 3.11, 3.12, and 3.13 show the source color image and the decompressed images obtained with three different compression rates by using the *YUV* *F*-transform method.

In Fig. 3.14 the trend of the PSNR with the compression rate is plotted and obtained by using the DCT, (*RGB*) *F*-transform, *YUV* *F*-transform, and JPEG compression method, respectively. Figure 3.14 shows that the *YUV* *F*-transform method improves the performance of the *F*-transform compression method applied in the *RGB* space.

Fig. 3.10 Original image

Fig. 3.11 F-tr decoded
image ($\rho = 0.14$)

3.5 Multilevel Fuzzy Transform Image Compression

In [8] the authors propose an extension of the F-transforms method (for short, MF-transforms) based on a multilevel decomposition of the image. The source image is compressed via the direct F-transform and decompressed via the inverse F-transform. Then, the error image, given by the difference between the decompressed and the source images, is constructed as well. This process is iterated by considering the error image as the new source image. The algorithm ends when a predetermined condition is reached

In Fig. 3.15 we schematize this multilevel architecture.

Fig. 3.12 *F*-tr decoded
image ($\rho = 0.28$)

Fig. 3.13 *F*-tr decoded
image ($\rho = 0.44$)

The reconstructed image is given by the sum of all the images decoded at any level. In the example of Fig. 3.15, if $I^{(1)}, I^{(2)}, I^{(3)}$ are the decoded images in the three levels, the final reconstructed image is given by $I_{\mathrm{REC}} = I^{(1)} + I^{(2)} + I^{(3)}$.

The main advantage of the multilevel *F*-transform algorithm is the ability to control the quality of the reconstructed image; this feature makes this method usable even in contexts in which it is necessary to limit and control the loss of information in the image.

In [8] three end iteration criteria were applied:

- the PSNR of the reconstructed image at the level h is greater than a prefixed threshold $\mathrm{PSNR_{th}}$. In this case, the quality of the reconstructed image obtained is already acceptable;

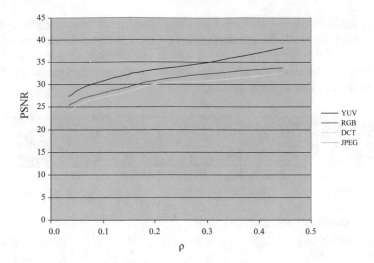

Fig. 3.14 PSNR trend obtained by coding/decoding color images with different methods

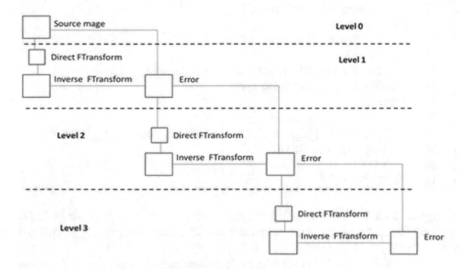

Fig. 3.15 Scheme of the multilevel image compression algorithm

- the difference between the PSNR at the hth level and the PSNR at the $(h-1)$th level is less than a difference threshold DPSNR$_{\text{th}}$. The algorithm stops because the contribution to the improvement of the image quality obtainable in the subsequent iterations will be of little significance;
- the process has reached a maximum number of iterations $h_{\max}$.

The multilevel F-transform image compression algorithm is shown below. Initially the variable I_0 contains the source image I_{source}. In any cycle the direct F-transform

I_D of I_0 and its inverse F-transform I are calculated. The reconstructed image at the hth level I_{REC} is calculated adding the inverse F-transform I to the reconstructed image calculated in the previous cycle. The algorithm ends if at least one of the three previous conditions is reached. The output of the algorithm is the set of direct F-transforms calculated in any cycle.

Algorithm: Multilevel F-transform Image Compression	
Input:	*Source image Isource*
Output:	*Direct F-transform in each level*
1	Set the threshold values $PSNR_{th}$ and $DPSNR_{th}$ and the max number of iterations
2	h:= 1
3	stopIteration = FALSE
4	$I_0 := I_{SOURCE}$
5	$I_{REC} := 0$
6	PSNRprev:= 0
7	WHILE stopIteration = TRUE
8	$I_D :=$ Direct FTransform of I_0
9	Restore I_D
10	I:= Inverse FTransform of I_D
11	$I_{REC} := I + I_{REC}$
12	Calculate the PSNR value comparing I_{REC} and I_{SOURCE}
13	DPSNR = PSNR - PSNRprev
14	h:= h + 1
15	$I_0 := I_0 - I$ // error image
16	IF (PSNR • $PSNR_{th}$) OR (DPSNR • $DPSNR_{th}$) OR (h > h_{max}) THEN
17	stopIteration = TRUE
18	END IF
19	END WHILE

Figures 3.16, 3.17, 3.18, 3.19, 3.20, and 3.21 show the results obtained in [8] applying this method to the original image 256×256 image Leopard (Fig. 3.16) and setting $PSNR_{th}$ to 26 and $DPSNR_{th}$ to 0.1.

The final image (Fig. 3.17) is given by the reconstructed image obtained in level 3, showing a PSNR greater than $PSNR_{th}$. The other figures show the reconstructed image and the error image obtained in levels 1 and 2.

Table 3.2 shows the PSNR and the PSNR difference obtained for the image reconstructed at any level. The PSNR obtained for the final reconstructed image (level 3) is greater than the threshold $PSNR_{th}$ equal to 26.

Fig. 3.16 Original image

Fig. 3.17 Final (level 3)
decoded image

3.6   Fuzzy Transform-Based Methods for Coding/Decoding Videos

A motion is a sequence of frames given by images with the same size and grayscale. In [10] the authors partition the frames of motion in intra-frames (for short, I-frames), given by original frames and predictive frames (for short, P-frames), with pixel values very close to the ones of the preceding I-frame; thus a P-frame can be obtained as the previous I-frame plus a difference frame, called Δ-frame. This schematization allows to compress with different compression rates, ρ_I and ρ_P with $\rho_I > \rho_P$, I-frames and P-frames, respectively. In fact, an I-frame can be compressed strongly than a P-frame, by compressing strongly a difference frame, called Δ-frame, representing the difference of information with respect to the previous P-frame, as this Δ-frame has a

Fig. 3.18 Level 1 decoded
image

Fig. 3.19 Level 1 error

low quantity of information and hence it can be coded with a low compression rate. In Fig. 3.24 we schematize a video partitioned into I-frames and P-frames (Fig. 3.22).

In [9] the authors classify the frames in P-frames and F-frames by applying a similarity index based on the Lukasiewicz t-norm.

We consider a frame F given by an $N \times M$ gray image with L gray levels. Now suppose that F is a P-frame. Let G be a successive frame. To classify G as an I-frame or a new P-frame, we normalize F and G considering them as fuzzy relations and apply the Lukasiewicz residuum operator.

The residuum operator $\rightarrow$ of a t-norm t is a binary operation in [0, 1] implementing the logic implication. If u, v, w are fuzzy sets or real numbers, we have that

$$\text{utv} \leq w \text{ if and only if } u \leq (v \rightarrow_t w) \tag{3.17}$$

Fig. 3.20 Level 2 decoded
Image

Fig. 3.21 Level 2 error

If t is the Lukasiewicz t-norm, we obtain

$$v \to_L w = \min(1 - v + w, 1) \tag{3.18}$$

The similarity index applied in [9] between the two fuzzy relations F and G is given by

$$\mathrm{Sim}_L(F, G) = \frac{\sum_{i=1}^{M} \sum_{j=1}^{N} \min\{F(i, j) \to_L G(i, j), G(i, j) \to_L F(i, j)\}}{M \times N} \tag{3.19}$$

Fig. 3.22 A video given by a sequence of *I*-frames and *P*-fames

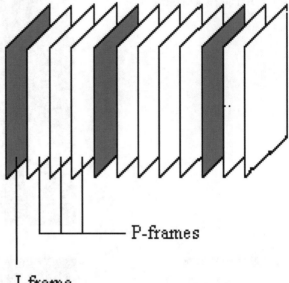

P-frames

I-frame

Fig. 3.23 Original first *I*-frame

Table 3.2 PSNR obtained at each level for the image Leopard

Level	PSNR	PSNR difference
1	24.675	–
2	25.709	1.034
3	26.181	0.472

If the similarity between the *I*-frame F and the frame G measured with the similarity index (3.19) is greater than a fixed threshold, then G is classified as a *P*-frame. In this case, given the Δ-frame D, where $D(i, j) = (F(i, j) - G(I, j) + 1)/2$, we can compress F with the compress rate ρ_I and D with the compression rate ρ_P, reconstructing G by the formula:

Fig. 3.24 Original first
P-frame

$$G'(i, j) = \frac{\max\{0, F'(i, j) - 2D'(i, j) + 1\}}{\max\{1, F'(i, j) - 2D'(i, j) + 1\}} \tag{3.20}$$

where G' is the reconstructed frame of G and F' and G' are the decoded frame, respectively, of F and D. In [9] the Fuzzy Relation Equation image compression algorithm is used to compress the frames. The first frame is classified as a *P*-frame F, then the similarity index with the next frame G is calculated. If this index is greater than the prefixed threshold, the delta frame D of G is compressed with the compression rate ρ_P, reconstructing G by (3.20). Conversely, if the similarity index is lower than the threshold, G is the new *P*-frame and the next frame is compared with it. This process ends when the last frame has been classified and compressed.

In [4–6] the *F*-transform method is applied to compress the video frames in color videos; the authors show that by compressing the frames with the *F*-transform method, the quality of the reconstructed frames is better with respect to the ones reconstructed via furry relation equation compression method.

Below we schematize in the pseudocode the video compression algorithm applied to compress a gray video. For a color video, this algorithm must be executed in any band.

Algorithm: F-transform *Video compression*
Input: Source Video
Output: *Compressed video*

Set Threshold, $\bullet_p, \bullet_I$
F:= Frame(1)
Classify F as an I-frame
Compress F via F-transform with compression rate $\bullet_I$
Store the compressed image of F
for each frame G
   Calculate Sim_L(F,G) via (3.19)
   if Sim_L(F,G) • Threshold *then*
      Classify G as a P-frame
      D:= (F-G+1)/2
      Compress D via F-transform with compression rate $\bullet_p$
      Store the compressed image of D
   else
      Classify G as an I-frame
      Compress G via F-transform with compression rate $\bullet_I$
      Store the compressed image of G
      F:= G
   endif
Next

The decompressed video can be obtained decompressing each *I*-frame via the inverse *F*-transform and decompressing each *P*-frame decompressing the Δ-frame and applying the formula (3.20). In [4–6] the *F*-transform method is applied to compress the video frames in color videos; the authors carry out a preprocessing phase to set the value of the similarity threshold which is set as the maximal similarity value between *F* and *G* for which in any band the PSNR difference between the reconstructed image G compressed as an *I*-frame and the reconstructed image of *G* compressed as a *P*-frame is below a value *L*.

Fig. 3.25 The related
Δ-frame

Fig. 3.26 The reconstucted
P-frame

In Figs. 3.23, 3.24, 3.25, and 3.26 are shown the first I-frame, the successive *P*-frame, the related Δ-frame, and the reconstructed *P*-frame for the color image composed by (about) 100 frames. In the preprocessing phase, a threshold value of about 0.19 has been determined.

References

1. Di Martino, F., Loia, V., & Sessa, S. (2003a). A method for coding/decoding images by using fuzzy relation equations. In T. Bilgic, B. De Baets & O. Kaynak (Eds.), *Fuzzy sets and systems—IFSA 2003*. Lecture Notes in Artificial Intelligence (Vol. 2715, pp. 436–441). Berlin: Springer.
2. Di Martino, F., Loia V., & Sessa S. (2003b). A method in the compression/decompression of images using fuzzy equations and fuzzy similarities. In *Proceedings of the 10th IFSA World Congress* (pp. 524–527). Istanbul, Turkey.
3. Di Martino, F., Loia, V., Perfilieva, I., & Sessa, S. (2008). An image coding/decoding method based on direct and inverse fuzzy transforms. *Fuzzy Sets and Systems, 48*(1), 110–131.
4. Di Martino, F., Loia, V., & Sessa S. (2010a). Fuzzy transforms for compression and decompression of color videos. *Information Sciences, 180*, 3914-3931.
5. Di Martino, F., Loia V., & Sessa S. (2010b). A segmentation method for images compressed by fuzzy transforms. *Fuzzy Sets and Systems, 161*(1), 56–74
6. Di Martino, F., Loia, V., & Sessa S. (2010c). Fuzzy transforms method and attribute dependency in data analysis. *Information Sciences, 180*(4), 493–505
7. Di Martino, F., Perfilieva I, & Sessa S. (2017). First order fuzzy transform for images compression. *Journal of Signal and Information Processing, 8*, 178–194.
8. Di Martino, F., & Sessa S. (2018). Multi-level fuzzy transforms image compression. *Journal of Ambient Intelligence and Humanized Computing*, 1–12. https://doi.org/10.1007/s12652-018-0971-4.
9. Loia, V., & Sessa, S. (2005). Fuzzy relation equations for coding/decoding processes of images and videos. *Information Sciences, 171*, 145–172.
10. Nobuhara, H., Hirota, K. Pedrycz W., & Sessa S. (2006). A motion compression/reconstruction method based on max T-norm composite fuzzy relational equations. *Information Sciences, 176*(17), 2526–2552.
11. Nobuhara, H., Pedrycz, W., & Hirota, K. (2005). Relational image compression: optimizations through the design of fuzzy coders and YUV color space. *Soft Computing, 9*(6), 471–479.
12. Nobuhara, H., Hirota, K., Di Martino, F., Pedrycz, W., & Sessa, S. (2005). Fuzzy relation equations for compression/decompression processes of colour images in the RGB and YUV colour spaces. *Fuzzy Optimization and Decision Making, 4*(3), 235–246.

13. Pennebaker, W. B., & Mitchell, J. L. (1991). *JPEG: Still Image Data Compression Standard* (638 pp). Springer. ISBN: 978-0442012724.
14. Perfilieva, I. (2006). Fuzzy transforms: Theory and applications. *Fuzzy Sets and Systems, 157*(8), 993–1023.

Chapter 4
Fuzzy Transform Technique for Image Autofocus

Image autofocus is an automatism applied to adjust the camera lens to obtain focus on the subject, and can mean the difference between a sharp photo and a missed opportunity. Despite a seemingly simple goal—sharpness at the focus point—the inner workings of how a camera focuses are unfortunately not straightforward.

The autofocus (AF) systems are of two types:

- *active systems*, measure the distance from the subject using measuring instruments based on ultrasounds or laser beams, and apply the focus adjustment on the optical system. These systems are electronically controlled; the distance data are processed by an electronic device that activates a mechanical or electro-mechanical system to adapt the focus to this data;
- *passive systems*, do not use devices to measure the distance from the subject, but find the optimal focus applying phase detection or contrast detection algorithms.

A problem of the active systems is that it is difficult to measure distances through transparent surfaces, since most of these surfaces reflect sound waves or infrared light. Conversely, passive systems can focus subject placed behind transparent surfaces, but they are ineffective in low light conditions and all other low contrast cases.

In the following we focus on passive autofocus methods, in particular on contrast detection methods. We will analyze in detail some passive AF contrast detection methods based on the use of the F-transform.

4.1 Passive Image Autofocus Techniques

Passive autofocus methods (PAF), commonly used on single-lens reflex cameras, devaluate the optimal distance from the subject via comparative analysis of the image itself seeking for the best focus. The PAF methods are classified into

© Springer Nature Switzerland AG 2020
F. Di Martino and S. Sessa, *Fuzzy Transforms for Image Processing and Data Analysis*,
https://doi.org/10.1007/978-3-030-44613-0_4

- Phase Detection AutoFocus (PDAF), in which the image in the camera is split into two images, by comparing how the two images diverge on the sensor and hence by finding the exact point at which the two images are in phase;
- Contrast Detection AutoFocus (CDAF) in which the contrast between the pixels is analyzed: the better the image is in-focus, the greater the difference in intensity between pixels is.

CDAF methods are generally slower than PDAF, but PDAF is far more accurate at focusing on still subjects. Both methods can be ineffective in low light.

Generally, a sharpness function is used for detecting the contrast in an image. This function is calculated on each captured image and reaches its optimal value at the in-focus image.

There is no need to calibrate your lens as there are no mechanical focusing errors. However, they often have problems in low light. The user must be also sure that they are focusing on the correct point (e.g. focusing on the subject, not in background).

4.2 Passive Image Autofocus: Contrast Detection Measures

CDAF algorithms are used in contrast autofocus systems for finding the contrast peak in focus curve, corresponding to the maximum of a sharpness function.

Let I be an $N \times M$ gray image and $I(x, y)$ the intensity of the pixel at the coordinate (x, y). A CDAF algorithm uses a specific index as a sharpness function to measure the contrast in the image: most of the in-focus image is given by the image with the highest value of this contrast index. Furthermore, a good CDAF algorithm must correctly distinguish out-focus images from images with focus on a subject.

Some well-known CDAF indices used in the literature are shown below [1].

- Variance and Normalized Variance

In [2], the contrast of the image is measured by computing the variance index given by

$$V = \frac{1}{M \cdot N} \sum_{x=1}^{M} \sum_{y=1}^{N} (I(x, y) - \mu)^2 \qquad (4.1)$$

where μ is the mean luminance of the image, defined as

$$\mu = \frac{1}{M \cdot N} \sum_{x=1}^{M} \sum_{y=1}^{N} I(x, y) \qquad (4.2)$$

In order to normalize the variance in different images, the following normalized variance can be used:

$$Vn = \frac{V}{\mu} \tag{4.3}$$

– Squared Gradient

Gradient-based index is based on the luminance gradients of an image: they consider a first-order difference to evaluate the gradient in the image.

The Squared Gradient index, proposed in [3], is based on the concept that larger gradients exert more influence on the image contrast and it is given by the sum of squared differences:

$$SG = \sum_{x=-1}^{M} \sum_{y=1}^{N} (I(x+1, y) - I(x, y))^2 \tag{4.4}$$

with the condition that $(I(x+1, y) - I(x, y))^2 \geq \theta$, where θ is a fixed gradient threshold.

– Tenenbaum Gradient

This index was proposed in [4, 5] calculates the magnitude of the 2D spatial gradients in the focused region by using the Sobel operator to convolute in the vertical and horizontal directions.

The gradient x and y components are given, respectively, by

$$G_x = \sum_{s=-1}^{1} \sum_{t=-1}^{1} S_x(s+1, t+1) \cdot I(x+s, y+t) \tag{4.5}$$

$$G_y = \sum_{s=-1}^{1} \sum_{t=-1}^{1} S_y(s+1, t+1) \cdot I(x+s, y+t) \tag{4.6}$$

where S_x and S_y are the 3×3 Sobel operators:

$$S_x = \begin{vmatrix} -1 & 0 & 1 \\ -2 & 0 & 2 \\ -1 & 0 & 1 \end{vmatrix} \quad S_y = \begin{vmatrix} 1 & 2 & 1 \\ 0 & 0 & 0 \\ -1 & -2 & -1 \end{vmatrix} \tag{4.7}$$

The Tenenbaum Gradient index is obtained by computing the sum of the squares of the values of the gradient in the two directions and summing it to all the pixels:

$$T = \sum_{x=-1}^{M} \sum_{y=1}^{N} (G_x(x, y))^2 + (G_y(x, y))^2 \tag{4.8}$$

– Sum Modified Laplacian

The Laplacian filter considers a second-order difference for evaluating the second-order derivative of the image intensity. The Sum Modified Laplacian index (SML) was proposed in [6] calculated by computing the absolute values of the convolution of the image with the Laplacian operators in the two directions. These absolute values are given as

$$\left| \frac{\partial^2 I(x, y)}{\partial x^2} \right| = |2I(x, y) - I(x - 1, y) - I(x + 1, y)| \tag{4.9}$$

$$\left| \frac{\partial^2 I(x, y)}{\partial y^2} \right| = |2I(x, y) - I(x, y - 1) - I(x, y + 1)| \tag{4.10}$$

Then the Sum Modified Laplacian (SML) index is given as

$$SML = \sum_{x=-1}^{M} \sum_{y=1}^{N} \left| \frac{\partial^2 I(x, y)}{\partial x^2} \right| + \left| \frac{\partial^2 I(x, y)}{\partial y^2} \right| \tag{4.11}$$

- Energy of Laplacian

Energy of Laplacian index (EoL) was proposed in [6, 7]. It differs from the SML in that the image is convolved with a 3×3 mask and the sum of the squares of the convolution results are used to measure the image contrast. A typical mask used to calculate the EoL index is the following matrix:

$$L = \begin{vmatrix} -1 & -4 & -1 \\ -4 & 20 & -4 \\ -1 & -4 & -1 \end{vmatrix} \tag{4.12}$$

The convoluted image is given by

$$C(x, y) = \sum_{s=-1}^{1} \sum_{t=-1}^{1} L(s + 1, t + 1) \cdot I(x + s, y + t) \tag{4.13}$$

Then the EoL index is given as

$$EoL = \sum_{x=-1}^{M} \sum_{y=1}^{N} (C(x, y))^2 \tag{4.14}$$

- Modified Entropy

Modified Entropy index (ME) was proposed in [8]. It is based on the entropy concept, considering the fact that the histogram of a focused image can point out more information than the histogram of a defocused image. Moreover, it is robust to the small irregular movements of the camera and to the presence of noise in the image. The following convoluted image is considered:

$$CE(x, y) = \frac{1}{8} \sum_{s=-1}^{1} \sum_{t=-1}^{1} |I(x, y) - I(x + s, y + t)| \tag{4.15}$$

And the ME index is given as

$$ME = \frac{1}{M \cdot N} \sum_{x=-1}^{M} \sum_{y=1}^{N} (CE(x, y) \cdot \log(CE(x, y) + 1) \tag{4.16}$$

- Contrast Measurement Adaptive to Noise (CMAN)

The Contrast Measurement Adaptive to Noise (CMAN) index was proposed in [9]. It is based on the same idea of the SML index but it is more robust to noise. The convoluted image is the same one that was used in the ME index, given by (4.15). The CMAN index is defined as

$$CMAN = \frac{1}{M \cdot N} \sum_{x=-1}^{M} \sum_{y=1}^{N} CE(x, y) \cdot \left(\sqrt[n]{CE(x, y) + 1} - 1 \right) \tag{4.17}$$

where the value of the parameter n is calculated by considering the number m of local maximums having values greater than a predefined threshold(s) T. The formula for calculating the parameter n is given as

$$n = \begin{cases} 1 \text{ if } m \leq T_1 \\ 2 \text{ if } T_1 < m \leq T_2 \\ 3 \text{ if } m > T_2 \end{cases} \tag{4.18}$$

when m is less or equal to the parameter T_1, the influence of the noise is low, and it does not affect the localization of the best focused image. If m is between T_1 and T_2, the noise adds a little bias to the contrast measure. If m is greater than T_2, the influence of the noise is high, so the best focused image is difficult to be located.

In [1], comparison tests are performed between various sharpness functions applied on image sequences representing different lens positions of the camera to measure the best image contrast. The results of these tests show that the best performances were obtained by using the Tenengrad and CMAN indices, but both Tenengrad and CMAN fail to detect the best focused image if the images are strongly influenced by noise. Then to improve the performances in the presence of noise in the image, frequency-based contrast measurements are required.

To improve the CDAF algorithm performances to focusing images in the presence of noise, some authors apply algorithms based on frequency measures. In [10] a focus measure in the S-transform domain that is based on the energy of high-frequency components is applied. In [11], 2014) a frequency measure based on the Discrete

Cosine Transform is proposed. In [12] an entropy sharpness function based on the discrete curvelet transform components is used for obtaining the in-focus image from a set of color microscopic images.

4.3 Direct Fuzzy Transforms Applied for Passive Image Autofocus

In order to obtain a CDAF algorithm fast and robust to the presence of noise, Roh et al. [13] propose a variation of the variance index based on the direct F-transform: they introduce the concept of fuzzy subspace of the image, that is the image is segmented in fuzzy subspaces based on the luminosity of the pixels. For segmenting the image, a set of two fuzzy clustering algorithms are used (see Chap. 6 for a discussion of fuzzy clustering image segmentation techniques). When an image is contaminated by noise, a local variance index measured in these subspaces is more suitable than the variance index on the overall image for evaluating the image contrast. In [13] the authors compute this local variance by using the unidimensional direct F-transform algorithm of the intensity function.

A fuzzy clustering algorithm is applied to the pixel values of the image: the center of a cluster is interpreted as the mean gray level of the corresponding fuzzy subregion. In Fig. 4.1b, we show the gray level frequency histogram of the image Lena (Fig. 4.1a). In Fig. 4.1c we show the centers of any cluster and the corresponding membership degree function (the activation), obtained in [13] by setting $n = 3$ clusters.

The centers of the clusters and the partition matrix are used to obtain the local variance values of any subregion.

Let $\mathbf{v}_1,...,\mathbf{v}_n$ be the centers of the n clusters, let $\mathbf{x}_j, j = 1, ..., N \times M$ be the jth pattern, corresponding to the jth pixel, and let $u_{ij}, i = 1, ..., n, j = 1, ..., N \times M$ be the membership degree of the jth pixel to the ith cluster. Then a fuzzy partition $\{A_1,...,A_n\}$ is created, defined as

$$A_i(\mathbf{x}_j) = \begin{cases} u_{ij} & \text{if } \mathbf{x_j} \notin \{\mathbf{v_1}, \ldots, \mathbf{v_n}\} \\ 1 & \text{if } \mathbf{x_j} = \mathbf{v}_i \\ 0 & \text{if } \mathbf{x_j} \in \{\mathbf{v}_1, \ldots, \mathbf{v}_{i-1}, \mathbf{v}_{i+1}, \ldots, \mathbf{v}_n\} \end{cases} \tag{4.19}$$

The set $\{A_1,...,A_n\}$ constitutes a uniform fuzzy partition of the pixel space. Moreover, the data are sufficiently dense with respect to this fuzzy partition because a constraint imposed in the fuzzy clustering algorithms requires that no cluster be empty, thus at least a pixel must belong to a cluster with a membership degree greater than 0.

In [13] the ith direct F-transform component constructed by using the basic functions (4.19) is interpreted as a weighted mean pixel intensity in the ith fuzzy subspace. It is given by

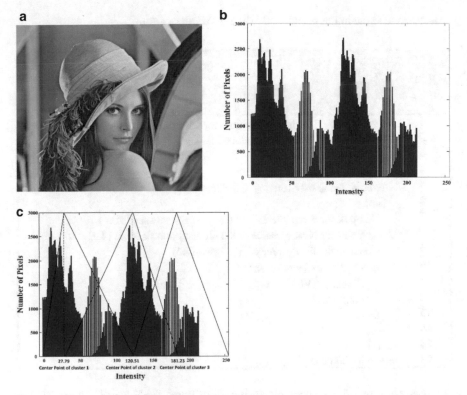

Fig. 4.1 **a** Source image, **b** histogram of the intensity values, **c** cluster centers on the histogram (by [13])

$$\mu_i = \frac{\sum_{j=1}^{N} I(x_j) A_i(x_j)}{\sum_{j=1}^{N} A_i(x_j)} \tag{4.20}$$

where $I(x_j)$ is the gray level of the jth pixel. The variance index in the ith fuzzy subspace is defined as

$$\text{VFS}_l = \frac{\sum_{j=1}^{N} \left[I(x_j) - \mu_l \right]^2 \cdot A_l(x_j)}{\sum_{j=1}^{N} A_l(x_j)} \quad l = 1, \ldots, n \tag{4.21}$$

The final variance used in [13] for measuring the contrast of the image is given by

$$\text{VFS} = \sum_{j=1}^{n} \text{VFS}_i \tag{4.22}$$

It is used to find the best focused image.

Below we show in the pseudocode the CDAF algorithm.

Algorithm: F-transform image fuzzy subspace Variance CDAF	
Input:	*Set of images* $I^{(1)}, ..., I^{(K)}$
	Number of clusters n
Output:	*In-focus image*
1	VFSmax := 0 // inizializzazione varianza più alta
2	ImInFocus ;= $I^{(1)}$ // inizializzazione immagine in focus
3	*for* each image $I^{(k)}$ $k = 1, ..., K$
4	Partition the image in n clusters
5	Execute the fuzzy clustering algorithm
6	*for* i = 1 to n
7	Compute the mean intensity of the ith fuzzy subspace via (4.20)
8	Compute the local variance of the ith fuzzy subspace via (4.21)
9	Compute the final variance VFS$^{(k)}$ via (4.22)
10	*if* (VFS$^{(k)}$ > VFSmax) *then*
11	VFSmax := VFS$^{(k)}$
12	ImInFocus := $I^{(k)}$
13	*end if*
14	*next* i
15	*next* k
16	*return* ImInFocus

A combination of two fuzzy clustering algorithms, the Fuzzy C-Means (FCM) algorithm [14] and the Conditional FCM (CFCM) algorithm [15], is used in order to detect the clusters forming the fuzzy subspaces.

The authors perform comparative tests with other CDAF methods, showing that such variance discriminates between in-focus and out-focus images better than methods based on the Tenenbaum, Sum of Modified Laplace, Modified Entropy, and Variance indices.

The comparisons are performed normalizing in the interval [0, 1] the contrast index calculated in any image.

Let o_h be an index value computed by using a CDAF algorithm analyzing the hth image in a set of t images. Its normalized value is given by

$$\bar{o}_h = \frac{\left(o_h - \min_{s=1,...,t} (o_s) \right)}{\left(\max_{s=1,...,t} (o_s) - \min_{s=1,...,t} (o_s) \right)} \quad h = 1, ..., t \tag{4.23}$$

In Fig. 4.2a–c we show a color image with three different subjects in focus. Figure 4.2d shows the out-focus image.

In Table 4.1 the results obtained in [13] are shown. The most in-focus image obtained with the F-transform Variance (VFS) is Fig. 4.2c. The most in-focus image

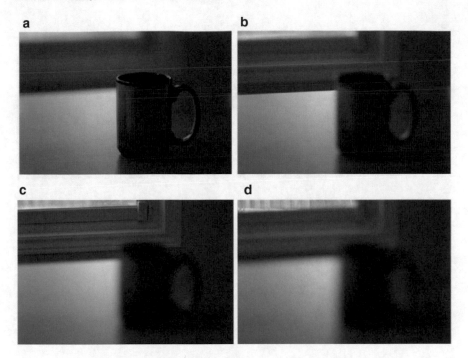

Fig. 4.2 **a** Cup in-focus, **b** Table in-focus, **c** Windows in-focus, **d**. Out-focus image

Table 4.1 Comparison results for Fig. 4.2 in [13]

Image	VFS	T	SML	V	ME
4.2a	0.41	0.11	0.03	0.27	0.43
4.2b	1.00	0.00	0.00	0.00	0.24
4.2c	0.62	1.00	1.00	0.54	1.00
4.2d	0.00	0.69	0.54	1.00	0.00

obtained by using the Tenenbaum (T), Sum of Modified Laplace (SML), Variance (V), and Modified Entropy (ME) contrast indices is Fig. 4.2c. Finally, the most focused image obtained using the Variance index is Fig. 4.2d, which is really the out-focus image. Only the F-transform Variance correctly identifies the out-focus image since the normalized value obtained for Fig. 4.2d is 0, unlike the other contrast indices.

To improve the performances of the CDAF algorithm in [13, 17], apply a robust variation of the FCM algorithm called Fuzzy Generalized Fuzzy C-Means (FGFCM) proposed in [18] and described in Sect. 6.5. Furthermore, in order to evaluate the optimal number of fuzzy subregions in the image, the authors propose a preprocessing phase in which the Partition Coefficient and Exponential Separation (PCAES) validity index [19] is calculated for different values of the number of clusters.

Fig. 4.3 **a** Vegetation in-focus, **b** forward leaf in-focus, **c** out-focus image

Figure 4.3a, b shows two images with two different subjects in-focus; Fig. 4.3c is the out-focus image.

In Table 4.2 we show the results obtained by comparing eight normalized indices obtained for the images of Fig. 4.3, where the VFSC index is the F-transform Variance index obtained in [16, 17] applying the FGFCM fuzzy clustering algorithm.

The results in Table 4.2 show that only the ME, CMAN, VFS, and VFSC indices give correctly the image in Fig. 4.3 completely out-focus.

Figure 4.4 shows the percentage of cases in which out-focus has been correctly detected applying each of the eight indices on an image dataset given by 100 sets of color images. Each image set is given by images focused on different objects of which only one is out-focus.

The best performances are given by using the VFSC index. Only using the VFS and the VFSC indices gives the percentage of correctly detected out-focus images over 90%.

Table 4.2 Comparison results for Fig. 4.3 in [17]

Image	T	SML	EoL	ME	CMAN	V	VFS	VFSC
4.3a	0.50	0.33	0.42	0.66	1.00	0.37	1.00	1.00
4.3b	0.00	1.00	0.00	1.00	0.56	1.00	0.65	0.78
4.3c	1.00	0.44	1.00	0.00	0.00	0.44	0.00	0.00

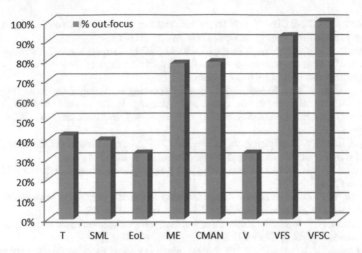

Fig. 4.4 Percentage of correctly detected out-focus images

The benefits of the F-transform block-wise fragile image watermarking algorithm consists in terms of source image archiving and execution times because it is more efficient than the classical F-transform watermarking algorithm (that is, the image is not divided into blocks) and it represents an optimal trade-off between accuracy and precision in the detection of tampered zones and execution times and memory allocation.

References

1. Xu, X., Zhang, X., Li, S., Liu, X., Wang, X., & Tang, J. (2014). A comparison of contrast measurements in passive autofocus systems for low contrast images. *Multimedia Tools Appl, 69*, 139–156.
2. Groen, F. C. A., Young, I. T., & Ligthart, G. (1985). A comparison of different autofocus algorithms. *Cytometry, 6*, 81–91.
3. Santos, S. A., Solórzano, C. O., Vaquero, J. J., Peña, J. N., Malpica, N., & Pozo, F. (1997). Evaluation of autofocus functions in molecular cytogenetic analysis. *Journal of Microscopy, 188*, 264–272.
4. Krotkov, E. (1987). Focusing. *International Journal of Computer Vision, 1*, 223–237.
5. Yeo, T., Jayasooriah, S. O., & Sinniah, R. (1993). Autofocusing for tissue microscopy. *Image and Vision Computing, 11*, 629–639.
6. Nayar, S. K., & Nakagawa, Y. (1994). Shape from focus. *IEEE Transactions on Pattern Analysis and Machine Intelligence, 16*, 824–831.
7. Subbarao, M., Choi, T. S., & Nikzad, A. (1993). Focusing techniques. *Optical Engineering, 32*(11), 2824–2836.
8. Tang, J., & Xu, X. (2009). An automatic focus algorithm for still and video camera applications using a new contrast measure. In: *MIPPR 2009. Proceedings SPIE 7498, Remote Sensing and GIS Data Processing and Other Applications*. https://doi.org/10.1117/12.833926.
9. Xu, X., Wang, Y., Tang, J., Zhang, X., & Liu, X. (2011). Robust automatic focus algorithm for low contrast images using a new contrast measure. *Sensors, 11*, 8281–8294.

10. Mahmood, M. T., & Choi, T. S. (2010). Focus measure based on the energy of high-frequency components in the S transform. *Optics Letters, 35,* 1272–1274.
11. Zhang, Z., Liu Y., Tang X., & Zhang M. (2014). Robust sharpness metrics using reorganized DCT coefficients for auto-focus application. In: *Proceedings of 12th Asian Conference on Computer Vision* (pp. 172–187). Singapore.
12. Dogan H., Ayas S., & Ekinci M. (2015). Auto-focusing with multi focus color image fusion based on curvelet transform on microscopic imaging. In *9th International Conference on Electrical and Electronics Engineering* (pp. 237–240). https://doi.org/10.1109/eleco.2015. 7394468.
13. Roh, S. B., Oh, S. K., Pedrycz, W., & Seo, K. (2016). Development of autofocusing algorithm based on fuzzy transform. *Fuzzy Sets and Systems, 288,* 129–144.
14. Bezdek, J. C. (1981). *Pattern recognition with fuzzy objective function algorithms.* New York: Plenum Press.
15. Pedrycz, W. (1998). Conditional fuzzy clustering in the design of radial basis function neural networks. *IEEE Transactions on Neural Networks, 9*(4), 601–612.
16. Di Martino, F., & Sessa, S. (2019). Complete image fusion method based on fuzzy transforms. *Soft Computing, 23*(6), 2113–2123.
17. Di Martino, F., & Sessa, S. (2019). Passive image autofocus by using direct fuzzy transform. *International Journal of Computational Science and Engineering, 20*(2). https://doi.org/10. 1504/ijcse.2017.1001188 (in press).
18. Chen, S. C., & Zhang, D. Q. (2004). Robust image segmentation using FCM with spatial constraints based on new kernel-induced distance measure. *IEEE Transactions on Systems, Man and Cybernetics, Part B, 34*(4), 1907–1916.
19. Wu, K. L., & Yang, M. S. (2005). A fuzzy validity index for fuzzy clustering. *Pattern Recognition Letters, 26,* 1275–1291.

Chapter 5
Fuzzy Transform for Image Fusion and Edge Detection

In this chapter we discuss F-transform based techniques for image fusion and edge detection. Image fusion is an important image processing technique with applications in many fields: it produces a single image from a set of source images in order to convey into it the parts of the original images containing the most significant information useful for the specific analysis. For example, the fused image can contain the less undistorted parts of the original images, or it contains different objects focused on multiple photos of the same scene.

Edge detection is a technique applied to detect pixel value local gradients in the image in order to detect points in the image in which image brightness changes rapidly: these points are typically organized into a set of curved line segments termed edges. Edge detection is a fundamental tool in image processing, machine vision, and computer vision, particularly in the areas of feature detection and feature extraction.

5.1 Image Fusion Concept

Image fusion aims at the integration of various complementary image data into a single, new image with the best possible quality with respect to the specific problem faced.

Let u be the ideal image we intend to obtain applying a fusion algorithm and c_1, ..., c_K be the acquired input images. We define the relation between each c_k, $k = 1$, 2, ..., K, and u expressed by [1]:

$$c_k(i, j) = P_k(u(i, j)) + e_k(i, j) \quad k = 1, \ldots, K \tag{5.1}$$

where P_k is an unknown operator describing the image degradation, and e_k is an additive random noise. The problem of fusion consists in finding an image û such

© Springer Nature Switzerland AG 2020
F. Di Martino and S. Sessa, *Fuzzy Transforms for Image Processing and Data Analysis*,
https://doi.org/10.1007/978-3-030-44613-0_5

that it is close to u and it is better (in terms of a chosen quality) than any of $c_1, \ldots, c_K$.

A variety of different fusion algorithms exist in the literature with specific adaptation for particular problems. The main characteristics of an image fusion algorithm are the following:

- extraction of the useful information from the source images (for example, extraction of the focused objects in any source image c_i);
- no introduction of artifacts in the output image;
- robustness to the imperfections in the original images.

We focus our attention on Multi-scale Decomposition (MSD) fusion techniques as the MSD techniques capture and provide information on local contrast changes, that contain extremely useful information for the human observer.

The most known MSD image fusion method is based on the pyramid transform. An image pyramid structure [2] is a multi-scale representation of the image in which the image is continuously blurred and subsampled with respect to the one at the previous level: as shown in Fig. (5.1), at each level the image resolution is halved and compared to one of the previous levels.

The pyramid transforms image fusion algorithms merge the levels of the pyramids of the source images, and construct the fused image by taking the inverse pyramid transform. The application of the merging operator is called the fusion rule. Several types of pyramid methods are used for image fusion: one of the most known types is the Laplacian pyramid [2, 3], in which the image at a level is constructed as a low-pass filter of the image in the previous level and a strength measure is applied to decide from which source which pixel contributes at each specific sample location. Another well-known pyramid image fusion technique is the Wavelet Transform [4] which uses the multi-scale Discrete Wavelet decomposition technique [5] in which a Discrete Wavelet Transform (shortly, DWT) creates a decomposition of the image in two horizontal and two vertical reduced images by a low-pass (L) and a high-pass (H) filters, respectively. Components at the same levels coming from different images are merged and the Inverse Wavelet Transform operator is applied to obtain the fused image. In Fig. 5.2 we schematize the DWT image fusion process: in the decomposition process, the two source images are broken down into three levels. The figure below shows the structure by levels produced by the decomposition process.

5.2 Image Decomposition via F-Transforms

Some authors propose MSD image fusion methods based on F-transform [6–11].

In [9] two image fusion methods based on F-transforms are proposed. In the first method called Simple Algorithm (shortly, SA), the decomposition process ends at the first level. In the second method, called Complete Algorithm (shortly, CA), the decomposition process ends to a level which is defined a priori.

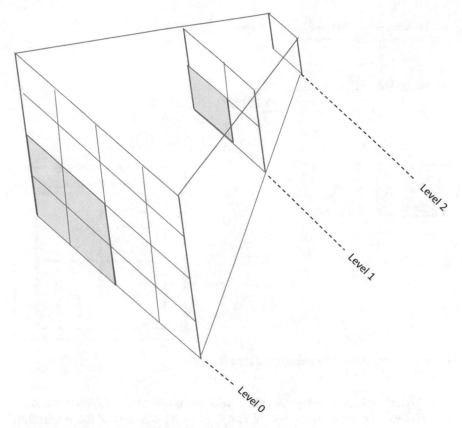

Fig. 5.1 Examples of an image pyramid multi-scale decomposition

These methods perform the decomposition of the image by applying the bi-dimensional F-transform. We consider an image u of size $N \times M$, compressed via the bi-dimensional F-transform in the $n \times m$ image with components:

$$F_{kl} = \frac{\sum_{j=1}^{M} \sum_{i=1}^{N} u(i, j) A_k(i) B_l(j)}{\sum_{j=1}^{M} \sum_{i=1}^{N} A_k(i) B_l(j)} \quad k = 1, \ldots, n \quad l = 1, \ldots, m \qquad (5.2)$$

where $A_k, k = 1, \ldots, n$ and $B_l, l = 1, \ldots, m$ are the basic functions. The decompressed image is given by

$$u_{nm}(i, j) = \sum_{k=1}^{n} \sum_{l=1}^{m} F_{kl} A_k(i) B_l(j) \quad i = 1, \ldots, N \quad j = 1, \ldots, M \qquad (5.3)$$

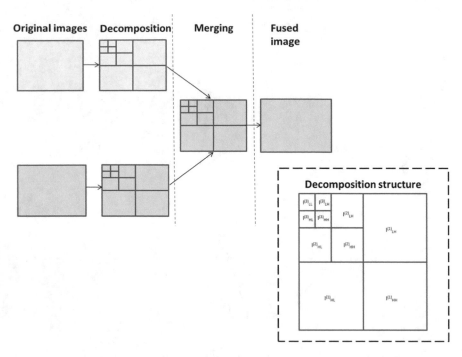

Fig. 5.2   Schema of the DWT image fusion process

In [9] an F-transform image decomposition process for image fusion is proposed. By (5.1) we can consider the source image given by the sum of the F-transform decompressed image [3, 5] and an error e_{NM} as

$$u = u_{nm} + e_{NM} \tag{5.4}$$

where, for brevity, we put $u = u(i, j)$. Equation (5.4) is called the *one-level decomposition* of the original image u. Since the image can contain several types of degradation, the error e_{NM} can be not negligible and then one-level will not be enough for the decomposition. Henceforth e_{NM} is substituted with its inverse F-transform with $n < n_1 \leq N$ and $m < m_1 \leq M$, obtaining that

$$u = u_{nm} + e_{n_1 m_1} + e^{(1)} \tag{5.5}$$

which is the second-level decomposition of u. Again, if the function $e^{(1)}$ is not negligible, we can replace it with its inverse F-transform $e_{n_2 m_2}$ with $n < n_1 \leq n_2 \leq N$ and $m < m_1 \leq m_2 \leq M$, thus obtaining

$$e^{(1)} = e_{n_2 m_2} + e^{(2)} \tag{5.6}$$

Obtaining the third-level decomposition of u, we have that

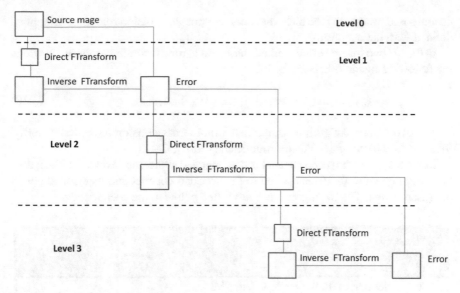

Fig. 5.3 Schema of the F-transform image decomposition process

$$u = u_{nm} + e_{n_1m_1} + e_{n_2m_2} + e^{(2)} \tag{5.7}$$

Iterating by $(p - 1)$ times this process we obtain the pth level decomposition of u as

$$u = u_{nm} + e_{n_1m_1} + e_{n_2m_2} + \cdots + e_{n_{p-1}m_{p-1}} + e^{(p-1)} \tag{5.8}$$

where $n < n_1 \le n_2 \le \cdots \le n_{p-1} \le N, m < m_1 \le m_2 \le \cdots \le m_{p-1} \le M$ and the error e^h is defined via the recursive formula

$$e^{(h)} = e^{(h-1)} - e_{n_h,m_h} \quad h = 2, \ldots, p - 1 \tag{5.9}$$

The source image can be reconstructed by [5, 8] unless an error $e^{(p-1)}$is obtained. In Fig. 5.3 we schematize the F-transform decomposition process, considering three levels.

5.3   F-Transform Image Fusion Algorithms: The CA, SA, and ESA Algorithms

Let $c_1, \ldots, c_K$ be K image channels $N \times M$ with various types of degradation; our goal is to recognize the undistorted parts of any channel fusing them in the fused image. A fusion rule must be applied to the components at the same level of the decomposed images to fuse them into only one image. A fusion rule must be applied defined with the operator.

In [9, 12] two image fusion methods based on F-transforms are proposed in which the following fusion rule is applied:

$$\xi(x_1, \ldots, x_K) = \max(|x_1|, \ldots, |x_K|) \quad x_1, \ldots, x_K \in \Re \qquad (5.10)$$

The two proposed algorithms are called Simple Image Fusion Algorithm (shortly, SA) and Complete Image Fusion Algorithm (shortly, CA).

In the SA algorithm each channel is decomposed at the one-level as in (5.4); then the fusion operator (5.10) is applied to the decoded channel and its error to obtain the fused image. The SA algorithm is described below in the pseudocode.

Algorithm: SA		
Input:	$N \times M$ image channels $c_1, \ldots, c_K$	
Output:	*Fused image c*	
1	Set n,m with $0 < n \leq N$, $0 < m \leq M$.	
2	*for* k := 1 to K	
3	Apply (5.2) and (5.3) to obtain the inverse F-transform c_{knm}	
4	Calculate the error $e_{knm} := c_k - c_{knm}$	
5	*next* k	
6	Apply (5.10) by obtaining $c_{nm} = \xi\left(c_{1nm}, c_{2nm}, \ldots, c_{Knm}\right)$	
7	Apply (5.10) by obtaining $e_{nm} = \xi\left(e_{1nm}, e_{2nm}, \ldots, e_{Knm}\right)$	
8	Construct the fused image $c = c_{nm} + e_{NM}$.	

The CA algorithm is similar to the SA algorithm, except for the fact that each image is decomposed to a level $p > 1$. The quality of the fused image is better than the one obtained by using the CA algorithm, but the CPU times of CA are much higher the higher the number of levels in the composition. The CA algorithm is described below in the pseudocode.

Algorithm: CA	
Input:	$N \times M$ *image channels* $c_1, ..., c_K$
	Number of cycles p
Output:	*Fused image c*

1	Set n,m with $0 < n \leq N$, $0 < m \leq M$.
2	*for* k := 1 to K
3	Apply (5.2) and (5.3) to obtain the inverse F-transform c_{knm}
4	Calculate the error $e_{knm} := c_k - c_{knm}$
5	*next* k
6	Apply (5.10) by obtaining $c_{nm} = \xi\left(c_{1_{nm}}, c_{2_{nm}}, ..., c_{K_{nm}}\right)$
7	$c := c_{nm}$
8	*for* h = 2 to p-1
9	Set a fuzzy partition formed by $n_h \times m_h$ basic functions $A_k \times B_l$, $k = n_1, ..., n_{p-1}$, $l = m_1, ..., m_{p-1}$, being $n < n_1 \leq n_2 \leq ... \leq n_{p-1} \leq N$, $m < m_1 \leq m_2 \leq .. \leq m_{p-1} \leq M$
10	Apply (5.9)) to each image channel by obtaining $e_1^{(h)} := e_1^{(h-1)} - e_{1_{n_h, m_h}}, ..., e_K^{(h)} := e_K^{(h-1)} - e_{K_{n_h, m_h}}$
11	Apply (5.10) by obtaining $e_{n_h m_h} := \xi\left(e_{1_{n_h, m_h}}, e_{2_{n_h m_h}}, ..., e_{K_{n_h m_h}}\right)$
12	
13	$c := c + e_{n_h m_h}$
14	*next* h
15	Apply (5.10) by obtaining $e^{(p-1)} := \xi\left(e_1^{(p-1)}, e_2^{(p-1)}, ... e_K^{(p-1)}\right)$
16	Construct the fused image $c := c + e^{(p-1)}$

To achieve a compromise between the reconstructed fused image quality and CPU times, proposed a variation of the SA called Extended Simple Algorithm (shortly, ESA). In ESA another decomposition level is performed with respect to the one-level decomposition of SA in order to reduce image fusion artifacts as ghosts and lakes effects. The fused image is given by a weighted sum of the source images where the weights are calculated by computing the differences $e_k^{(1)} = e_{k_{NM}} - e_{k_{nm}}$, $k = 1, ..., K$. Below we schematize in the pseudocode the ESA algorithm.

Algorithm: ESA	
Input:	$N \times M$ *image channels* $c_1, ..., c_K$
Output:	*Fused image c*
1	Set n,m with $0 < n \leq N$, $0 < m \leq M$.
	w:= 0
2	*for* k := 1 to K
3	Apply (5.2) and (5.3) to obtain the inverse F-transform c_{knm}
4	Calculate the error $e_{knm} := c_k - c_{knm}$
5	Compute the weight $\quad w_k = e_k^{(1)} = e_{kNM} - e_{knm}$
6	w:= w+w_k
7	$c_k := c_{knm} w_k$
8	*next* k
9	Apply (5.10) by obtaining $c = \xi(c_1, c_2, ..., c_K)/w$

5.4   The CCA Algorithm

In [11] a variation of the CA algorithm, called Controlled Complete Algorithm (shortly, CCA) is proposed. In CCA the depth of the decomposition process is not set a priori, but is determined during the decomposition process by applying a criterion on the error image. The decomposition process is stopped at the hth level when the maximum of the absolute value of the error of any source image is below a specified threshold ε.

Formally, if $e_k^{(h)} Z$ is the error difference obtained at the hth level by decomposing the kth source image, with $k = 1, ..., K$, this criterion is given by

$$\max\left(\left|e_1^{(h)}\right|, \left|e_2^{(h)}\right|, ..., \left|e_K^{(h)}\right|\right) \leq \varepsilon \quad 2 \leq h \leq p \tag{5.11}$$

Below we schematize in the pseudocode the CCA algorithm.

Algorithm: CCA	
Input:	$N \times M$ image channels $c_1, ..., c_K$
	Maximum number of cycles p
	Error difference threshold ε
Output:	Fused image c

1   Set n,m with $0 < n \leq N$, $0 < m \leq M$.

2   *for* k := 1 to K

3   Apply (5.2) and (5.3) to obtain the inverse F-transform c_{knm}

4   Calculate the error $e_{knm} := c_k - c_{knm}$

5   *next* k

6   Apply (5.10) by obtaining $c_{nm} = \xi\left(c_{1_{nm}}, c_{2_{nm}}, ..., c_{K_{nm}}\right)$

7 $c := c_{nm}$

8   h:=2

9   StopCriterion: = FALSE   // criterion used to stop the cycle

10   *while*(h < p) AND (StopCriterion = *false*)

11   Set a fuzzy partition formed by $n_h \times m_h$ basic functions $A_k \times B_l$,
k= $n_1, ..., n_{p-1}$, l = $m_1, ..., m_{p-1}$, being $n < n_1 \leq n_2 \leq ... \leq n_{p-1} \leq N$, $m < m_1 \leq m_2 \leq .. \leq m_{p-1} \leq M$

12   Apply (5.9) to each image channel by obtaining
$e_1^{(h)} := e_1^{(h-1)} - e_{1_{n_h, m_h}}, ..., \quad e_K^{(h)} := e_K^{(h-1)} - e_{K_{n_h, m_h}}$

13   Apply (5.10) by obtaining
$e_{n_h m_h} := \xi\left(e_{1_{n_h, m_h}}, e_{2_{n_h m_h}}, ..., e_{K_{n_h m_h}}\right)$

14   *if* $\max\left(\left|e_1^{(h)}\right|, \left|e_2^{(h)}\right|, ..., \left|e_K^{(h)}\right|\right) \leq \varepsilon$ *then*

15   put $e_1 := e_{1_{n_h m_h}}, e_2 := e_{2_{n_h m_h}}, ..., e_K := e_{K_{n_h m_h}}$

16 $c := c + e_{n_h m_h}$

17   h:=h+1

18   *end if*

19   *end while*

20   Apply (5.10) by obtaining $e^{(h)} := \xi\left(e_1^{(h)}, e_2^{(h)}, ... e_K^{(h)}\right)$

21   Construct the fused image $c := c + e^{(h)}$

The CCA algorithm, in addition to represent a trade-off between the quality of the fused image and the execution times with respect to the ESA method, has the advantage of verifying that the quality of the fused image is acceptable. On the other hand, a critical point of the CCA algorithm is given by the choice of the error difference threshold ε (Fig. 5.4).

Figures 5.5 and 5.6 show two source images of the 270×205 gray image *Leaves*. In the first channel the leaves closer to the camera are focused, in the second image the most distant leaves are focused as well. In Figs. 5.7, 5.8, 5.9, and 5.10, we show the fused images obtained in [11] by using the SA, CA, ESA and CCA algorithms, respectively, by setting $\varepsilon = 0.03$ and $p = 4$ and using a compression rate

Fig. 5.4 Leaves—Channel 1

Fig. 5.5 Leaves—Channel 2

Fig. 5.6 SA fused image

Fig. 5.7 CA fused image

Fig. 5.8 ESA fused image

Fig. 5.9 CCA fused image

Fig. 5.10 Image Lenna

Table 5.1 Quality indexes for the fused image Leaves [11]

Quality index	Image fusion algorithm			
	SA	CA	ESA	CCA
MSE	68.15	25.67	26.05	20.50
PSNR	29.80	34.04	33.97	35.01

Table 5.2 Execution time indexes for the fused image Leaves [11]

Execution time index	Image fusion algorithm			
	SA	CA	ESA	CCA
CPU time (s)	0.31	12.54	4.32	7.05

$\rho = 0.25$. The final level of decomposition obtained by applying the CCA algorithm is $h = p = 4$.

In Table 5.1, we show the MSE and PSNR indexes obtained. The best results are obtained by applying the CCA algorithm (PSNR $= 35.01$).

Table 5.2 shows the CPU times measured. As expected, the execution time of the CCA algorithm is greater than that related to the ESA algorithm, but less than that due to the CA algorithm, even if the level of decomposition is the same for both methods.

5.5 Edge Detection Concept

Edge detection is a fundamental image processing activity that detects pixels in an image at which the image brightness changes sharply; a set of adjoint pixels at which image brightness changes sharply form polylines is termed *edges*. Edge detection is used to detect and extract features in the image.

An edge detection algorithm (image detectors) finds edges evaluating gradient intensity and directions of the pixel.

Formally, if the image is represented as a continuous bi-dimensional function $f(x,y)$, its gradient is given by

$$\nabla f = \begin{bmatrix} g_x \\ g_y \end{bmatrix} = \begin{bmatrix} \dfrac{\partial f}{\partial x} \\ \dfrac{\partial f}{\partial y} \end{bmatrix} \tag{5.12}$$

The intensity and the direction of the gradient of the function f in the point (x, y) are given by

$$G(x, y) = \sqrt{g_x^2 + g_x^2}$$

$$\theta(x, y) = \tan^{-1}\left(\frac{g_x}{g_y}\right) \tag{5.13}$$

In the discrete case we can approximate the gradient of an image I in the point (x, y) as

$$g_x \approx \frac{I(x + 1, y) - I(x - 1, y)}{2}$$

$$g_y \approx \frac{I(x, y + 1) - I(x, y - 1)}{2} \tag{5.14}$$

Well-known image detectors are the Sobel and Prewitt algorithm [13] that uses a gradient operator and the Marr and Hildreth algorithm [14] that extracts the edge pixels using the second derivative of the Gaussian filter. The Haralick algorithm [15] detects the edges, finding the zero-crossings of the second directional taken in the direction of a nonzero gradient at the pixel's center.

One of the more widely used edge detectors is the Canny algorithm [16], in which an optimization process is used to detect the edges, The Canny algorithm can be synthesized in the following four steps [17]:

1. a Gaussian is applied to smooth the image reducing details and noise, so that image noise does not create false edges;
2. the intensity gradient and the direction gradient at each pixel are calculated;
3. if the intensity gradient at a pixel is larger than those at its two neighbors in the gradient direction, mark this pixel as an edge;
4. the weak edges are removed by applying two hysteresis thresholds—high (TH) and low (TL); if the gradient intensity is below the low threshold, the pixel is discharged: above the high threshold, the pixel is accepted as part of an edge. Between the two thresholds, the pixel is accepted only if it is adjoint to a previously accepted pixel. The use of two thresholds with hysteresis guarantees greater flexibility compared to the single threshold.

The choice of the three parameters used in the Canny algorithm (the standard deviation smooth of the Gaussian filter and the two hysteresis thresholds) can influence the performance of the results produced.

The blurring filter applied in the first step directly influences the results generated by the algorithm. Smaller filters produce less blur, and allow us to recognize sharper contours. Larger filters produce greater and are better at recognizing wider and more nuanced areas.

In addition, a threshold interval set at too high values can cause the loss of significant information, while a threshold set at too low values can cause the preservation of irrelevant information as disturbances, that will be interpreted as important elements of the image.

It is not possible to set generic Gaussian smoothing parameter and hysteresis thresholds interval that will be good for all images, and it is necessary to evaluate what parameter settings give satisfactory results for any image.

5.6 F^1-Transform Method for Edge Detection

In [18, 19] a Canny detector algorithm based on the F^1-transform is proposed. The idea presented is to obtain the gradient intensity and direction by using the F^1-transform technique in order to calculate the first partial derivative in the horizontal and vertical directions.

In this method the gradient components of the image I are obtained by the direct F^1-transform of the image.

Let I be an $N \times M$ gray image with $i = 1, ..., N$ and $j = 1, ..., M$. Let $A_k, k = 1, ..., n$ and $B_l, l = 1, ..., m$ be an h-uniform triangular-shaped fuzzy partitions with $n < N$ and $m < M$.

By [19, Lemma 8], the direct F^1-transform components C_{kl}^{10} and c_{kl}^{01} can be approximated with, respectively:

$$c_{kl}^{10} = \frac{\sum_{j=1}^{M} \sum_{i=1}^{N} I(i, j)(i - k) A_k(i) B_l(j)}{h^4} \tag{5.15}$$

$$c_{kl}^{01} = \frac{\sum_{j=1}^{M} \sum_{i=1}^{N} I(i, j)(j - l) A_k(i) B_l(j)}{h^4} \tag{5.16}$$

By [19, Theorem 10], C_{kl}^{10} approximate the partial derivative of I with respect to x in the node of coordinates (x_k, y_l):

$$c_{kl}^{10} \approx \frac{\partial I}{\partial x}(x_k, y_l) \tag{5.17}$$

Likewise, c_{kl}^{01} approximate the partial derivative of I with respect to y in the node of coordinates (x_k, y_l):

$$c_{kl}^{01} \approx \frac{\partial I}{\partial y}(x_k, y_l) \tag{5.18}$$

The partial derivative of I with respect to the x and y coordinates in the point (i, j) are obtained by calculating the inverse F^1-transform

$$\frac{\partial I}{\partial x}(i, j) \approx \sum_{k=1}^{n} \sum_{l=1}^{n} c_{kl}^{10} A_k(i) B_l(j) \tag{5.19}$$

$$\frac{\partial I}{\partial y}(i, j) \approx \sum_{k=1}^{n} \sum_{l=1}^{n} c_{kl}^{01} A_k(i) B_l(j) \tag{5.20}$$

In [19] steps 1 and 2 of the Canny detector are performed smoothing the image via the F^1-transform and calculating the gradient intensities and directions by using (5.19) and (5.20). Then, steps 3 and 4 are executed.

Below in the pseudocode, the F^1-transform Canny edge detection algorithm is schematized. The $N \times M$ gray image is coded in an $n \times m$ image with $0 < n \leq N$, $0 < m \leq M$ and decoded via the F^1-transform. The final result is an $N \times M$ binary image E, in which $E(i, j) = 1$ if the pixel (i, j) is an edge.

Algorithm: F transform Canny edge detector	
Input:	*N×M image I*
	Hysteresis thresholds TH and TL
Output:	*Edge matrix E*

1	Set n,m with $0 < n \leq N$, $0 < m \leq M$.
2	*for* k := 1 to n
3	*for* l := 1 to m
4	Calculate the direct F'-transform component c_{kl}^{10} via (5.15)
5	Calculate the direct F'-transform component c_{kl}^{01} via (5.16)
6	*next l*
7	*next* k
8	*for* i := 1 to N
9	*for* j := 1 to M
10	Calculate $\dfrac{\partial I}{\partial x}(i,j)$ via (5.19)
11	Calculate $\dfrac{\partial I}{\partial y}(i,j)$ via (5.20)
12	Calculate G(i,j) and θ (i,j) via (5.13)
13	*if* G(i,j) > TH *then*
14	E(i,j) := 1
15	*else*
16	*if* G(i,j) > TL *then*
17	E(i,j) := 0
18	*else*
19	E(i,j) = -1
20	*end if*
21	*end if*
22	*next* j
23	*next* i
24	*for* i := 1 to N
25	*for* j := 1 to M
26	*if* E(i,j) = 0 *then*
27	*if* G(i,j) is larger than those at its two neighbors in the gradient direction θ(i,j) *then*
28	E(i,j) := 1
	else
	E(i,j) := -1
29	*end if*
30	*end if*
31	*next* j
32	*next* i
33	*return* E

In Figs. 5.10, 5.11, 5.12, and 5.13, we show the 512×512 source image Lena, the edge obtained by using the Canny algorithm blurring the image with a 5×5 window and the edges obtained by applying the Canny F^1-transform algorithm, by using $T_L = 10$, $T_H = 20$, and $T_L = 10$, $T_H = 30$, respectively, [19].

Perfilieva et al. [19] proved that the F^1-transform provides edge results comparable with the ones obtained by using the Canny detector and, in addition, although both algorithms have a linear computational complexity, the complexity of the Canny algorithm depends on the value of a standard deviation σ used to blur the image.

Fig. 5.11 Canny detector edges on a 5×5 window

Fig. 5.12 F^1-tr edges ($T_L = 10$, $T_H = 20$)

Fig. 5.13 F^1-tr edges ($T_L = 10, T_H = 30$)

References

1. Sroubek, F., & Flusser, J. (2005). Fusion of blurred images. In L. Z. Blum, (Ed.), *Multi-sensor image fusion and its applications*. Signal Processing and Communications Series (pp. 423–438). San Francisco: CRC Press.
2. Burt, P. J. (1981). Fast filter transforms for image processing. *Computer Graphics and Image Processing, 16*(1), 20–51.
3. Adelson, E. H., Anderson, C. H., Bergen, J. R., Burt, J. R., & Ogden, J. (1984). Pyramid methods in image processing. *RCA Engineering, 29*(6), 33–41.
4. Pajares, G., & de la Cruz, J. M. (2004). A wavelet-based image fusion tutorial. *Pattern Recognition, 37*, 855–1872.
5. Mallat, S. G. (1989). A theory for multiresolution signal decomposition: the wavelet representation. *IEEE Transaction on Pattern Recognition and Machine Intelligence, 11*(7), 674–693.
6. Dankova, M., & Valasek, R. (2006). Full fuzzy transform and the problem of image fusion. *Journal of Electrical Engineering, 12*(2006), 82–84.
7. Perfilieva, I., Novàk, V., & Dvoràk, A. (2008). Fuzzy transforms in the analysis of data. *International Journal of Approximate Reasoning, 48*, 36–46.
8. Perfilieva, I., & Dankova, M. (2008). Image fusion on the basis of fuzzy transforms. In *Proceedings of the 8th International FLINS Conference on Computational Intelligence in Decision and Control*, Madrid (pp. 471–476).
9. Hodakova, P., Perfilieva, I., Dankova, M., & Vajgl, M. (2011). F-transform based image fusion. In O. Ukimura (Ed.), *Image fusion* (pp. 3–22). Rijeka: InTech. https://doi.org/10.5772/14504.
10. Manchanda, M., & Sharma, R. (2018). An improved multimodal medical image fusion algorithm based on fuzzy transform. *Journal of Visual Communication and Image Representation, 51*, 76–94.
11. Di Martino, F., & Sessa, S. (2019). Complete image fusion method based on fuzzy transforms. *Soft Computing, 23*(6), 2113–2123.
12. Perfilieva, I., Dankova, M., Hodakova, P., & Vajgl, M. (2010). The use of F-transform for image fusion algorithms. In *Proceedings of the International Conference of Soft Computing and Pattern Recognition, Cergy-Pontoise (F)* (pp. 472–477).
13. Rosenfeld, A., & Kak, A. (1982). *Digital picture processing* (2nd ed., 349 pp.). New York: Academic Press. ISBN: 978-0-12-597302-1.

14. Marr, D., & Hildreth, E. (1980). Theory of edge detection. *Proceedings of the Royal Society of London, 207*(1167), 187–217.
15. Haralick, R. M. (1984). Digital step edges from zero crossing of second directional derivatives. *IEEE Tranactions of Pattern Analysis and Machine Intelligence PAMI, 6*(1), 58–68.
16. Canny, J. A. (1986). Computational approach to edge detection. *IEEE Transactions on Pattern Analysis and Machine Intelligence, 8*(6), 679–698.
17. Ding, L., & Goshtasby, A. (1984). On the Canny edge detector. *Pattern Recognition, 34*(3), 721–725.
18. Perfilieva, I., Hodáková, P., & Hurtík, P. (2012). F^1-transform edge detector inspired by Canny's algorithm. In S. Greco, B. Bouchon-Meunier, G. Coletti, M. Fedrizzi, B. Matarazzo, & R. R. Yager (Eds.), *Advances on computational intelligence. IPMU 2012.* Communications in Computer and Information Science (Vol. 297). Berlin, Heidelberg: Springer.
19. Perfilieva, I., Hodáková, P., & Hurtik, P. (2016). Differentiation by the F-transform and application for edge detection. *Fuzzy Sets and Systems, 288*, 96–114.

Chapter 6
Fuzzy Transform for Image Segmentation

Image segmentation is a technique applied in image processing and computer vision to homogeneous detect regions of an image based on characteristics such as intensity, and texture. The segmentation refers to the process of partitioning an image into multiple disjoint regions of pixels homogeneous with respect to its characteristics in order to analyze and locate objects and boundaries in the image.

Many image segmentation algorithms were proposed in the literature: in this book we consider the two most famous types of image segmentation algorithms: threshold segmentation and clustering segmentation.

Threshold segmentation is one of the most commonly used segmentation techniques in region-based segmentation algorithm [1]. Our aim is to partition the pixels of the images in classes based on their gray-level values. A threshold algorithm searches the optimal gray-level thresholds to perform this partition.

Clustering-based segmentation algorithms apply a clustering algorithm to segment the pixels in the image space in order to partition the image dataset into a set of disjoint groups or clusters and a cluster contains pixels with similar characteristics.

6.1 Image Segmentation Concept

Image segmentation supports us to detect different objects in an image. The most natural way to segment objects in an image is to analyze the pixel values. For example, if we desire to distinguish objects with respect to the image background in the image in Fig. 6.1, we can analyze the pixel gray-level values, to detect the objects marking the pixels with gray-level values greater than a specified threshold or clearly distinguished from the values assumed by the pixels in the background.

The pixel values will be different for the objects and the image's background if there's a sharp contrast between them. To obtain the segmented binary image in Fig. 6.2 and detect the two geometric figures, we need to find the best value of this threshold. Generally, to detect n segments in an image with L gray levels it needs to set $n-1$ threshold values $T_1, \ldots, T_n$ with $0 < T_1, \ldots, T_n < L-1$. To set the n thresholds

© Springer Nature Switzerland AG 2020   81
F. Di Martino and S. Sessa, *Fuzzy Transforms for Image Processing and Data Analysis*,
https://doi.org/10.1007/978-3-030-44613-0_6

Fig. 6.1 Original image

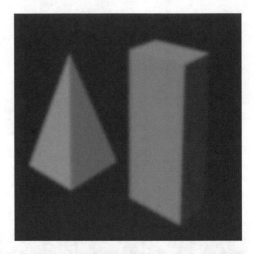

Fig. 6.2 Segmented image

is the aim of the image thresholding algorithms: the choice of the thresholds is the critical point of image thresholding method, since different values of the thresholds determine different pixel's classifications.

Another well-known technique applied in the image uses clustering methods to partition the image into clusters, called segments: an object in the image can be detected as an area of the image whose pixels belong to a specified segment. The pixels in a segment exhibits similar features as image intensity or texture. Partitive clustering algorithms are generally used to segment the image, setting the number of segments.

6.2 Image Thresholding—Fuzzy Entropy Maximization

Let I be an image with $N \times M$ pixels and L gray levels. In a bi-level thresholding, the image is partitioned in two classes by setting a threshold T between 0 and $L-1$.

The threshold T can be set by analyzing the histogram of the pixel gray level frequencies as in the example of Fig. 6.3 in which the threshold T is set in correspondence to a gray level position between distinct gray levels with high frequencies of pixels.

To set the optimal value of the threshold is a serious problem. A well-known algorithm applied to set this threshold is the Otsu method [2]; this optimization algorithm sets the threshold as that value which minimizes the intra-class variance and therefore it maximizes the inter-class variance between the two classes. It is based on the concept of increasing similarity between pixel frequencies with gray levels belonging to a class and separability between pixel frequencies with gray levels belonging to distinct classes. In [2] the threshold is selected by maximizing the inter-class variance but this method doesn't take into account spatial correlations between pixels. In addition, it is not robust with respect to noise, whose presence in the image produces an incorrect evaluation of the threshold.

In [3] an image thresholding method based on the maximization of an entropy function is proposed. Some authors use a fuzzy entropy function in multilevel image thresholding to find the thresholds $T_1,\ldots,T_r$.

Let $D = \{(i,j): i = 0,\ldots,M-1; j = 0,\ldots,N-1\}$ be the set of the pixels of an image I of sizes $N \times M$ and L be gray levels. Let $I(i,j)$ be the gray value of I at the position (i,j). We can decompose the set D in subsets:

$$D_k = \{(i, j) : I(i, j) = k, \quad (i, j) \in D\}, k = 0, 1, \ldots, L - 1 \qquad (6.1)$$

where the elements of the subset D_k are all the pixels of I with gray level k. The histogram of the image I is given by the set $H = \{h_0, h_1,\ldots, h_{L-1}\}$, where h_k is the

Fig. 6.3 Example of gray levels frequencies histogram

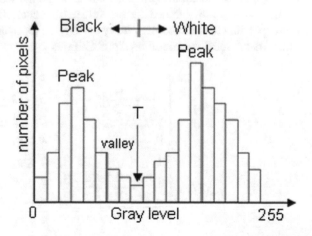

relative frequency of pixels with gray level equal to k, given by

$$h_k = \frac{n_k}{M \cdot N} \quad k = 0, 1, \ldots, L - 1 \tag{6.2}$$

and $n_k = \text{card } D_k$ is the number of pixels with gray level equal to k.

We can interpret h_k as the probability p_k that a pixel belongs to the subset D_k. Let $T_1, \ldots, T_r$ be the r selected thresholds and C_s the subset containing all the pixels with a gray value between T_{s-1} and T_s, with C_1 given by the pixels with gray level less or equal T_1 and C_{r+1} given by the pixels with gray level greater T_r and less or equal L, so we can do segmentation of the histogram of I into a set of $r + 1$ distinct classes. By Bayes theorem, the probability P_{ks} that a pixel, having gray level k, that belongs to the class C_s is given by

$$p_{ks} = h_k \cdot p_{s|k} \quad k = 0, 1, \ldots, L - 1, \quad s = 1, 2, \ldots, r + 1 \tag{6.3}$$

where $p_{s|k}$ is the conditional probability of belongness to C_s. Following [4], we can model the conditional probability $p_{s|k}$ via fuzzy sets set $\mu_s : k \in C_s \rightarrow \mu_s(k) \in [0, 1]$, $s = 1, \ldots, r + 1$. We introduce the fuzzy entropy index given by

$$e = \sum_{s=1}^{r+1} e_s \tag{6.4}$$

where

$$e_s = -\sum_{k=0}^{L-1} \frac{P_{ks}}{P_s} \cdot \log\left(\frac{P_{ks}}{P_s}\right) = -\sum_{k=0}^{L-1} \frac{P_k \cdot \mu_s(k)}{P_s} \cdot \log\left(\frac{P_k \cdot \mu_s(k)}{P_s}\right) \tag{6.5}$$

The thresholds are those that maximize the fuzzy entropy: to the maximum entropy corresponds the maximum information. The optimal values of the thresholds T_s^*, $s = 1, \ldots, r + 1$ are obtained for the values of k such that $\mu_s(k) = \mu_{s+1}(k) = 0.5$, corresponding to the maximum fuzzy entropy. In [5, 6] are used the following fuzzy numbers to model the conditional probabilities $p_{s|k}$:

$$\mu_1(k) = \begin{cases} 1 & \text{if } k \leq a_1 \\ 1 - \frac{(k-a_1)^2}{(c_1-a_1)\cdot(b_1-a_1)} & \text{if } a_1 < k \leq b_1 \\ \frac{(k-c_1)^2}{(c_1-a_1)\cdot(c_1-b_1)} & \text{if } b_1 < k \leq c_1 \\ 0 & \text{if } k > c_1 \end{cases}$$

$$\mu_s(k) = \begin{cases} 1 & \text{if } k \leq a_1 \\ 1 - \frac{(k-a_s)^2}{(c_s-a_s)\cdot(b_s-a_s)} & \text{if } a_s < k \leq b_s \\ \frac{(k-c_s)^2}{(c_s-a_s)\cdot(c_s-b_s)} & \text{if } b_s < k \leq c_s \\ 0 & \text{if } k > c_s \end{cases}$$

$$\mu_{r+1}(k) = \begin{cases} 0 & \text{if } k \leq a_{r+1} \\ \frac{(k-a_{r+1})^2}{(c_{r+1}-a_{r+1})\cdot(b_{r+1}-a_{r+1})} & \text{if } a_{r+1} < k \leq b_{r+1} \\ 1 - \frac{(k-a_{r+1})^2}{(c_{r+1}-a_{r+1})\cdot(c_{r+1}-b_{r+1})} & \text{if } b_{r+1} < k \leq c_{r+1} \\ 1 & \text{if } k > c_{r+1} \end{cases} \tag{6.6}$$

where $s = 2,\ldots, r$ and $0 < a_1 \leq b_1 \leq c_1 \leq \cdots \leq a_s \leq b_s \leq c_s \leq \cdots \leq a_r \leq b_r \leq c_r < L-1$. By using these fuzzy set, the optimal thresholds, $T_s^*, s = 1,\ldots, r+1$, are given by

$$T_s^* = \begin{cases} a_s + \sqrt{(c_s - a_s) \cdot \frac{(b_s - a_s)}{2}} & \text{if } \frac{(a_s + c_s)}{2} \leq b_s \leq c_s \\ c_s - \sqrt{(c_s - a_s) \cdot \frac{(c_s - b_s)}{2}} & \text{if } a_s \leq b_s \leq \frac{(a_s + c_s)}{2} \end{cases} \tag{6.7}$$

In Fig. 6.4 we show an example of membership functions (6.6) with four classes by considering $L = 256$, $a_1 = 10$, $b_1 = 35$, $c_1 = 60$, $a_2 = 89$, $b_2 = 130$, $c_2 = 160$, $a_3 = 180$, $b_3 = 200$, and $c_3 = 240$.

The optimal thresholds in this example are $T_1^* = 35$ (obtained imposing $\mu_1(k) = \mu_2(k) = 0.5$), $T_2^* = 127$ (obtained imposing $\mu_2(k) = \mu_3(k) = 0.5$), and .. (obtained imposing $\mu_3(k) = \mu_4(k) = 0.5$). For these values the fuzzy entropy (6.4) is maximum. The goal of the multilevel thresholding fuzzy entropy algorithm is to detect the values of the coefficients $a_s, b_s,$ and c_s with $a_s \leq b_s \leq c_s$ and $s = 1,\ldots, r$ that maximize the fuzzy entropy (6.4).

To explore the space of the $3r$ coefficients and detect the r thresholds, some bio-inspired heuristics and meta-heuristics algorithms such as Genetic Algorithm (GA), Firefly Algorithm (FA), Whale Optimization Algorithm (WOA), Gray Wolf Optimizer (GWO), and Particle Swarm Optimization (PSO) (see, for example, [7–9]) are used. A serious problem of these meta-heuristic algorithms is the increase of computational complexity with increasing image size.

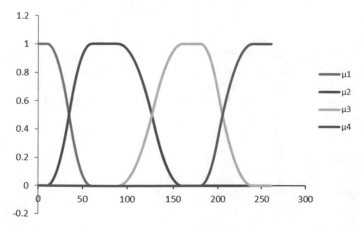

Fig. 6.4 Example with 4 fuzzy number ($r = 3$) in the form (6.6)

6.3   Fuzzy Transform Method for Image Thresholding

Di Martino and Sessa [10] propose a novel multilevel image thresholding method based on the bi-dimensional F-transform in which the Darwinian Chaotic PSO algorithm is applied to the image compressed via the direct bi-dimensional F-transform.

The Darwinian PSO algorithm is applied in [11, 12] for multilevel image threshold; these authors show that the performances are better than those obtained by using other meta-heuristic algorithms.

To overcome the drawback of getting trapped in local solution, Suresh and Lal [13] propose a variation of the Darwinian PSO algorithm, called the Chaotic Darwinian PSO algorithm, in which the initial positions of the particles are assigned by using a chaotic map. Then the authors apply this algorithm for multilevel image thresholding showing that this variation improves the performances of the Darwinian PSO algorithm.

The basic idea in [10] is to reduce the processing time necessary to determine the thresholds without affecting the quality of the results. The compression of the source image is realized via an iterative method in which the histograms obtained from the source and the compressed image are compared. Initially a compression rate ρ_0 is set; the bi-dimensional direct F-transform is calculated and the histograms of the source and compressed images are produced. Let I be a $N \times M$ source image with L gray levels and n_{0l} be the absolute frequency of pixels with gray level l, $l = 0,\dots,$ $L - 1$, and relative frequency $h_{0l} = \frac{n_{0l}}{N \cdot M}$. Let R be the compressed image of sizes $N_R \times M_R$, with $N_R < N$ and $M_R < M$ and $\rho = \frac{N_R \cdot M_R}{N \cdot M}$, obtained by compressing I via direct F-transform with a compression rate ρ, n_l the absolute frequency of pixels of R with a gray level l, and $h_l = \frac{n_l}{N_R \cdot M_R}$ its relative frequency. To measure the similarity between the two histograms a similarity index I_H is calculated as

$$I_H = \sum_{l=0}^{L-1} \min(h_l, h_{0l}) \tag{6.8}$$

If I_H is greater or equal to a specified threshold I_H^{Th}, then the process is iterated and a stronger compression (namely a lower compression rate) is used. The algorithm stops when I_H is lower than I_H^{Th}: the best compression rate is given by the previous compression rate used. Below we show the algorithm used to find the best compression rate.

Algorithm: FindBestCompRate	
Input:	$N \times M$ *source image I*
	Threshold histogram index I_H^{Th}
	Initial compression rate ρ_0
Output:	*Best compression rate* ρ_{BEST}

1	$\rho := \rho_0$
2	$\rho_{BEST} := 1$
3	Calculate the histogram of I
4	*while* $I_H \geq I_H^{Th}$
5	Calculate the compressed image R via the bi-dimensional direct F-transform
6	Calculate the histogram of R
7	Calculate the index I_H by (6.8)
8	*if* $I_H \geq I_H^{Th}$ THEN
9	$\rho_{BEST} := \rho$
10	Decrease ρ
11	*end if*
12	*end while*
13	*return* ρ_{BEST}

If $\rho_{BEST} = 1$, then the image cannot be compressed because even the minimum compression ($\rho = \rho_0$) produces a significant gray levels frequency histogram variation. In [10] a comparison of the F-transform Chaotic Darwinian PSO (FTR-CDPSO) with the Darwinian PSO (DPSO) and Chaotic Darwinian PSO (CDPSO) algorithms is performed. A logistic chaotic function is used and the initial number of swarms is set to 5 with 30 particles per swarm; the division of the image in blocks strategy is adopted in the F-transform image compression method. In the experiments, the number of thresholds varies from 3 to 6 and the PSNR and SSIM indexes comparing the original and the segmented image are calculated to measure the quality of the results and the CPU time is calculated to measure the processing times. In Table 6.1 we show the results obtained for the 512×512 source image Lena.

These results show that that the quality of the results obtained by using the FTR-CDPSO algorithm are comparable with the ones obtained by using CDPSO and better than that one obtained by using the DPSO algorithm. Moreover, the CPU time in the FTR-CDPSO algorithm is reduced by an average of over ½ of the CPU times measured by using the other two algorithms.

In Fig. 6.5 we show the segmented image of the 512×512 source image Lena via the FTR-CDPSO algorithm by setting 3, 4, 5, 6 thresholds, respectively.

Table 6.1 Comparison results for the image Lena

Thresholding method	Number of thresholds	Optimal thresholds	PSNR	SSIM	CPU time (s)
DPSO	3	60, 112, 175	19.76	0.812	30.33
	4	54, 97, 141, 178	23.07	0.828	56.23
	5	53, 98, 130, 155, 190	24.34	0.835	68.12
	6	51, 96, 117, 146, 158, 193	26.01	0.848	79.03
CDPSO	3	62, 115, 173	20.68	0.826	29.88
	4	53, 98, 136, 181	24.01	0.851	54.37
	5	50, 98, 125, 154, 188	25.19	0.873	67.05
	6	50, 93, 116, 145, 163, 195	26.75	0.887	76.73
FTR-CDPSO	3	64, 118, 170	20.59	0.824	10.18
	4	56, 100, 135, 182	23.89	0.847	18.72
	5	51, 97, 126, 152, 186	25.12	0.865	29.29
	6	49, 90, 114, 143, 161, 197	26.68	0.882	42.16

Threshold=3   Threshold=4

Threshold=5   Threshold=6

Fig. 6.5 Segmented images of Lena by using the FTR-CDPSO algorithm

6.4 Partitive Clustering Image Segmentation Algorithms

Partitive clustering algorithms are used in image segmentation to partition the pixels in the image in clusters, called segments. The most known partitive clustering algorithm is the K-means algorithm [14], in which the cluster prototypes are points in the Euclidean space of the features. At any iteration, a pattern is assigned to the nearest cluster.

Let $\mathbf{X} = \{x_1,\ldots, x_N\} \subset R^n$ be the dataset of the patterns given composed from N elements in the space of the n features. Let $\mathbf{V} = \{v_1,\ldots, v_k\} \subset R^n$ be the K clusters to be detected. The K-means algorithm is an iterative algorithm that minimizes the following objective function

$$J(\mathbf{X}, \mathbf{U}, \mathbf{V}) = \sum_{i=1}^{K}\sum_{j=1}^{N} h_{ij}d_{ij}^2 = \sum_{i=1}^{K}\sum_{j=1}^{N} h_{ij}\left\|x_j - v_i\right\|^2 \qquad (6.9)$$

where $h_{ij} = 1$ if the jth feature belongs to the ith cluster, 0 otherwise and $d_{ij} = \left\|x_j - v_i\right\|$ is the Euclidean distance between v_i and x_j. Initially the K clusters are assigned randomly, then each feature is assigned to the nearest cluster. In the next iteration, the cluster positions are re-calculated as the mean position of all the features belonging to the cluster, given by

$$v_i = \frac{1}{N_i}\sum_{j=1}^{N} h_{ij}x_j \qquad (6.10)$$

for $i = 1,\ldots, K$, where N_i is the number of patterns belonging to the ith cluster.

The process ends when no pattern has changed clusters or a maximum number of iterations has been completed. The result of the image segmentation K-means algorithm can be given by an image in which to any pixel is assigned a gray level value related to the identifier of the corresponding cluster: for example, if $K = 5$ and $L = 256$, considering an equal distance between two consecutive gray levels and assigning the gray level 0 to the pixels belonging to the first cluster and 255 pixels belonging to the last cluster, we obtain the five gray levels: 0, 63, 127, 191, 255. Below we show the image segmentation K-means algorithm.

Algorithm: image segmentation K-means	
Input:	*N×M source image I*
	Number of clusters K
	Max number of iterations max Iterations
Output:	*The segmented image*
1	Assign randomly the centers of the clusters
2	stopIteration:= *FALSE*
3	iter:= 0
4	*while* (stopIteration=*FALSE*)
5	Assign each feature to the nearest cluster
6	Calculate the cluster centers $\mathbf{v}_i$ i = 1,2,...,K by (6.10)
7	iter:= iter + 1
8	*if* (no pattern has changed cluster) OR (iter = maxIterations) *THEN*
9	stopIteration:= *TRUE*
10	*end if*
11	*end while*
12	Create the i-th segmented image
13	*return* the segmented image

The K-means algorithm has the advantage of being quite fast, as a few calculations are required. However, K-means is dependent on the initial number of clusters and easily falls into local optimum. In addition, when it is used in image segmentation, in which the patterns are the pixels of the image, this algorithm is not robust to the presence of noise in the image.

Some hybrid image segmentation method using K-means heuristic and meta-heuristic algorithms are proposed in the literature: in [15] a K-means hybrid model based on the dynamic PSO algorithm for image segmentation is proposed in order to improve the global search of K-means, in [16] a hybrid algorithm combining K-means with Firefly algorithm is proposed for brain image segmentation.

The K-means method does not allow to manage the uncertainty in the attribution of a pattern to a cluster. To manage this uncertainty, many image segmentation approaches use the well-known fuzzy C-means (for short, FCM) algorithm [17].

In the FCM algorithm, the jth pattern belongs to the ith cluster with a membership degree $u_{ij} \in [0, 1]$. If C is the number of cluster, the $C \times N$ matrix $\mathbf{U} = \begin{bmatrix} u_{ij} \end{bmatrix}$ is called *partition matrix*. In FCM the objective function is given by

$$J(\mathbf{X}, \mathbf{U}, \mathbf{V}) = \sum_{i=1}^{C} \sum_{j=1}^{N} u_{ij}^{m} d_{ij}^{2} = \sum_{i=1}^{C} \sum_{j=1}^{N} u_{ij}^{m} \left\| x_j - v_i \right\|^2 \qquad (6.11)$$

where d_{ij} is the Euclidean distance between the ith cluster and the jth pattern, the parameter $m \in [1, +\infty)$ is called *fuzzifier* which determines the degree of fuzziness of the partition and is generally equal to 2.

Moreover, the following constraints apply to FCM:

$$\sum_{i=1}^{C} u_{ij} = 1 \qquad \forall j \in \{1, \ldots, N\}$$

$$0 < \sum_{j=1}^{N} u_{ij} < N \ \forall i \in \{1, \ldots, C\} \tag{6.12}$$

Equation (6.12) requires that the sum of the membership degrees of a pattern to all the clusters is 1 and no cluster can be empty, in the sense that at least one pattern must belong to a cluster with membership degree greater than 0. By applying the Lagrange method on (6.11) we obtain

$$v_i = \frac{\sum_{j=1}^{N} u_{ij}^m x_j}{\sum_{j=1}^{N} u_{ij}^m} \tag{6.13}$$

$$u_{ij} = \frac{1}{\left(\sum_{k=1}^{c} \frac{d_{ij}^2}{d_{kj}^2}\right)^{\frac{2}{m-1}}} \tag{6.14}$$

for $i = 1, \ldots, C$ and $j = 1, \ldots, N$. Initially the membership degree values are assigned randomly and the cluster centers are calculated via (6.13) and at any iteration the membership degrees are re-calculated via (6.14). The process stops when a maximum number of iterations is reached or the following criterion holds:

$$\max\left\{\left|u_{ij}^{(s)} - u_{ij}^{(s-1)}\right| : (i, j) \in \{1, \ldots, C\} \times \{1, \ldots, N\}\right\} = \left|U^{(s)} - U^{(s-1)}\right| < \varepsilon \tag{6.15}$$

where $\varepsilon > 0$ is a parameter assigned a priori to stop the iteration process and $\mathbf{U}^{(s)} = \left[u_{ij}^{(s)}\right]$ is the partition matrix calculated at the sth step. The solution $(U^{(s)}, V^{(s)})$ converges to the local minima of the objective function (6.11).

Results of the image segmentation FCM are C segmented images. The ith segmented image is obtained assigning to the jth pixel the gray level approximating the value $(L-1)$ to u_{ij}: it is 0 if the membership degree of the jth pixel to the ith cluster is 0 and $L-1$ if this membership degree is 1. The FCM algorithm for image segmentation is shown below.

	Algorithm: image segmentation FCM
Input:	*N×M source image I*
	Number of clusters C
	Fuzzifier m
	Stop iteration criterion ε
	Max number of iterarions maxIterations
Output:	*The C segmented images*

1	Assign randomly the partition matrix components
2	Calculate the cluster centers $\mathbf{v}_i$ i = 1,2,...,C by (6.13)
3	stopIteration:= *FALSE*
4	iter:= 0
5	*while* (stopIteration=*FALSE*)
6	Calculate the partition matrix **U** by (6.14)
7	Calculate the cluster centers $\mathbf{v}_i$ i = 1,2,...,C by (6.13)
8	iter:= iter + 1
9	*if* ($\|U^{(s)} - U^{(s-1)}\| < \varepsilon$) OR (iter = maxIterations) *then*
10	stopIteration:= *TRUE*
11	*end if*
12	*end while*
13	*for* i = 1 to C
14	Create the i-th segmented image
15	*next* i
16	*return* the C segmented images

6.5   Extensions of FCM for Image Segmentation

As K-means algorithm, FCM needs to select the number of clusters a priori and is not robust with respect to the presence of noise in the image. In particular, FCM can be very sensitive to the presence of outliers and noises because it does not consider pixel's information of local spatial relations with the nearest pixels. Some researchers (see [18–20]) propose changes in the objective functions to consider relationships with the neighbor pixels. In [18], an additive term is added in the objective function (6.11) obtaining

$$J(\mathbf{X}, \mathbf{U}, \mathbf{V}) = \sum_{i=1}^{C} \sum_{j=1}^{N} u_{ij}^m \|x_j - v_i\|^2 - \frac{\alpha}{N} \sum_{i=1}^{C} \sum_{j=1}^{N} u_{ij}^m \sum_{k \in N_w} \|x_k - v_i\|^2 \quad (6.16)$$

where N_w is the number of neighbor pixels in a square window around the pixel x_j and α is a parameter controlling the contribution of the neighbor pixels.

To simplify the computational complexity of the algorithm in [20], the following variation of (6.16) is proposed:

$$J(\mathbf{X}, \mathbf{U}, \mathbf{V}) = \sum_{i=1}^{C} \sum_{j=1}^{N} u_{ij}^{m} \|x_j - v_i\|^2 - \frac{\alpha}{N} \sum_{i=1}^{C} \sum_{j=1}^{N} u_{ij}^{m} \|\tilde{x}_j - v_i\|^2 \qquad (6.17)$$

where $\tilde{x}_j$ is the mean or the median of the neighbor pixels in the window around the jth pixel. In [21] the authors proposed the EnFCM algorithm as a variation of the FCM algorithm in which the jth pixel x_j is transformed with the following:

$$\xi_j = \frac{1}{1+\alpha} \left(x_j + \frac{\alpha}{N} \sum_{k \in N_w} x_k \right) \qquad (6.18)$$

and the objective function (6.16) is replaced from the following formula:

$$J(X, U, V) = \sum_{i=1}^{C} \sum_{l=1}^{q} \gamma_l u_{il}^{m} (\xi_l - v_i)^2 \qquad (6.19)$$

where $q < N$ is the number of distinct gray level values and γ_l is the number of pixels in the transformed image with gray level l.

Applying the Lagrangian multiplicators to (6.19) we obtain

$$u_{il} = \frac{(\xi_l - v_i)^{-\frac{2}{m-1}}}{\sum_{k=1}^{C} (\xi_l - v_k)^{-\frac{2}{m-1}}} \qquad (6.20)$$

and

$$v_i = \frac{\sum_{l=1}^{q} \gamma_l u_{il}^{m} \xi_l}{\sum_{l=1}^{q} \gamma_l u_{il}^{m}} \qquad (6.21)$$

The parameter α balances the robustness to noises and the effectiveness of preserving the details of the image: the calculus of the best value for α requires a long computational time. Thus a fast variation of the EnFCM method, called FGFCM, is proposed in [19] to incorporate local spatial and gray level information together. The authors introduce a new parameter S_{jk} given from

$$S_{jk} = \begin{cases} S_{s\_jk} \cdot S_{g\_jk} & \text{if } k \neq j \\ 0 & \text{if } k = j \end{cases} \qquad (6.22)$$

where S_{s_jk} measures the influence of the kth pixel in the set of the neighbors to the jth pixel and decreases exponentially with the distance from this last pixel and S_{g_jk} is a measure of the local gray level similarity. The parameter S_{s_jk} is defined as

$$S_{s\_jk} = \exp\left(\frac{-\max(|p_j - p_k|, |q_j - q_k|)}{\lambda_s} \right) \qquad (6.23)$$

where p_j and q_j are the coordinates of the jth pixel and the parameter λ_s determines the spread of S_{s_jk}. The parameter S_{g_jk} is defined as

$$S_{g\_jk} = \exp\left(\frac{-\|x_j - x_k\|^2}{\lambda_g \cdot \sigma_{g\_j}^2}\right) \tag{6.24}$$

where x_j is the gray value of the jth pixel and λ_g determines the spread of S_{g_jk}. The parameter σ_{g_j} is a function of the local density surrounding the jth pixel defined as

$$\sigma_{g\_j} = \sqrt{\frac{\sum_{k \in N_w} \|x_j - x_k\|^2}{N}} \tag{6.25}$$

The more the local region surrounding the jth pixel is smoothed, the more the value of σ_{g_i} is small. FGFCM considers the following pixel value transformation as an alternative to (6.18):

$$\xi_j = \frac{\sum_{k \in N_w} S_{jk} x_k}{\sum_{k \in N_w} S_{jk}} \tag{6.26}$$

FGFCM can be summarized as follows:

Algorithm: FGFCM	
Input:	$N \times M$ source image I
	Number of clusters C
	Fuzzifier m
	Stop iteration criterion ε
	Max number of iterarions maxIterations
Output:	The C segmented images

1	Assign randomly the center of the clusters
2	Compute the grey level relashionship S_{jk} via (6.22)
3	Compute the linearly-weighted transformed image (6.26)
4	stopIteration:= *FALSE*
5	iter:= 0
6	*while* (stopIteration=*FALSE*)
7	Calculate the partition matrix **U** by (6.20)
8	Calculate the cluster centers $\mathbf{v}_i$ i = 1,2,...,C by (6.21)
9	iter:= iter + 1
10	*if* ($\|U^{(s)} - U^{(s-1)}\| < \varepsilon$) OR (iter = maxIterations) THEN
11	stopIteration:= TRUE
12	*end if*
13	*end while*
14	*for* i = 1 to C
15	Create the i-th segmented image
16	*next* i
17	*return* the C segmented images

6.6 F-Transform FGFCM Algorithm for Image Segmentation

F-transform can be applied to clustering image segmentation algorithms in order to reduce the computation times. In fact, we can compress the image by calculating the bi-dimensional direct F-transform and applying a compression rate so as to obtain an acceptable loss of information that does not compromise the results of the clustering algorithms. The clustering algorithm can be much faster because it is applied to the compressed image.

The problem to be solved is to determine the optimal compression rate in order to obtain the trade-off between the reduction of computation times and the quality of the results.

In [22] a hybrid algorithm is proposed in which the bi-dimensional F-transform is used for coding the source image and FGFCM is applied to extract the segmented images. To evaluate the optimal compression rate, the authors consider the following similarity index between the results obtained by using the source and the compressed image:

$$\text{Sim}_t(R_0, R_{f,\rho}) = \frac{1}{C} \sum_{i=1}^{C} \text{Sim}_t(R_o^i, R_{f,\rho}^i) \tag{6.27}$$

where R_o^i, $i = 1, 2, ..., C$, is the fuzzy relation corresponding to the ith segmented image obtained by applying the FGFCM algorithm to the source image and $R_{f,\rho}^i$, $i = 1, 2, ..., C$, is the fuzzy relation obtained by decoding the ith segmented image obtained by applying the FGFCM algorithm to the image compressed via F-transform with a compression rate ρ. The operator Sim_t is given by the following formula:

$$\text{Sim}_t(R_o^i, R_{f,\rho}^i)$$
$$= \frac{1}{M \times N} \sum_{h=1}^{M} \sum_{k=1}^{N} \min\{r_0^i(h, k) \to_t r_{f,\rho}^i(h, k), r_{f,\rho}^i(h, k) \to_t r_0^i(h, k)\} \tag{6.28}$$

where the symbol $\to_t$ denotes the residuum operator of a triangular norm t. The authors considers three residuum operators, corresponding to the t norm of Gödel, Goguen and Lukasiewicz, respectively, where

Gödel $a\, t\, b = \min(a, b)$ $a \to_t b = \begin{cases} 1 \text{ if } a \le b \\ b \text{ if } a > b \end{cases}$

Goguin $a\, t\, b = \min(a, b)$ $a \to_t b = \begin{cases} 1 & \text{if } a \le b \\ b/a \text{ if } a > b \end{cases}$

Lukasiewicz $a\, t\, b = \max(0, a + b - 1)$ $a \to_t b = \min(1, 1 - a + b)$ (6.29)

To evaluate the best compression rate, the source image is compressed with different compression rates: the best compression rate is obtained as the compression rate in which the similarity index is greatest or equal to a threshold θ.

Below the algorithm proposed in [22] is shown to evaluate the best compression rate.

Algorithm: searchBestCompressionRate	
Input:	$N \times M$ *source image I*
	Initial compression rate ρ_{IN}
	Threshold similarity θ
Output:	*Best compression rate* ρ_{Best}

1	Apply FGFCM to the source image
2	$\rho := \rho_{IN}$
3	$\rho_{Best} : \rho$
4	Compress the source image via F-transform
5	stopIteration:= *FALSE*
6	iter:= 0
7	*while* (stopIteration=*FALSE*)
8	Calculate the Similarity Index (6.27)
9	*If* $Sim_t(R_0, R_{f,\rho}) \geq \theta$ *then*
10	$\rho_{Best} : \rho$
11	Use a lower compression rate ρ
12	*else*
13	stopIteration:= *true*
14	*end if*
15	*end while*
16	*return* ρ_{Best}

The authors test this method to a sample of 100 gray 256×256 images. To set the similarity index threshold, a trend of the similarity index (6.28) varying the compression rate is plotted: the authors showed that this trend is similar by using all the three residuum operators. For this reason, they analyze the trend of the similarity index obtained by using the Lukasiewicz residuum. In Fig. 6.6 the trend of the similarity index obtained by applying the three residuum operators for the source image Lena is shown as well. The trend shows a plateau approximatively for $\rho > 0.25$, then the threshold value $\theta = 0.25$ is used.

To find the optimal number of clusters, the authors use a validity index to measuring the compactness of the clusters and the separation between clusters: the Partition Coefficient And Exponential Separation (for short, PCAES) validity index [23]; the optimal value of the number C of cluster is the one that maximizes the PCAES validity index.

In Fig. 6.7 we show the source image Lena and the image compressed via F-transform by using the compression rate $\rho = 0.25$

Figures 6.8, 6.9, and 6.10 show the segmented images obtained by applying the FGFCM algorithm to the original and the compressed image, setting the number of cluster C equal to 3 (Fig. 6.11).

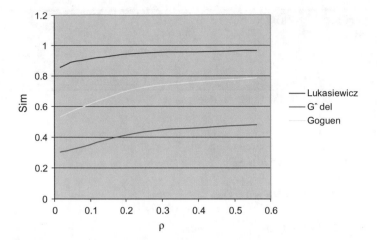

Fig. 6.6 Trend of similarity varying the compression rate for Lena [22]

Fig. 6.7 The source image Lena and the image compressed with $\rho = 0.25$

Fig. 6.8 First segmented images obtained for image Lena and the compressed one ($\rho = 0.25$)

Fig. 6.9 s segmented images obtained for image Lena and the compressed one ($\rho = 0.25$)

Fig. 6.10 Third segmented images obtained for image Lena and the compressed one ($\rho = 0.25$)

Fig. 6.11 Fourth segmented images obtained for image Lena and the compressed one ($\rho = 0.25$)

References

1. Aha, D. W. (Ed.) (1997). *Lazy learning*, p. 432. Norwell (MA): Kluwer Academic Publishers. ISBN: 0792345843.
2. Otsu, N. (1979). A threshold selection using grey level histograms. *IEEE Transactions on Systems Man Cybernetics, 9,* 62–69.
3. Kapur, J. N., Sahoo, P. K., & Wong, A. K. C. (1985). A new method for gray-level picture thresholding using the entropy of the histogram. *Computer Vision, Graphics, and Image Processing, 29,* 273–285. https://doi.org/10.1016/0734-189X(85)901252.
4. Chao, Y., Dai, M., Chen, K., Chen, P., & Zhang, Z. (2016). Fuzzy entropy based multi-level image thresholding using modified gravitational search algorithm. In *IEEE International Conference Industrial Technology (ICIT)*, pp. 752–757. Taipei (Taiwan): IEEE Publisher. https://doi.org/10.1109/icit.2016.7474845.
5. Tao, W. B., Tian, J. W., & Liu, J. (2003). Image segmentation by three-level thresholding basic on maximum fuzzy entropy and genetic algorithm. *Pattern Recognition Letters, 24*(16), 3069–3078. https://doi.org/10.1016/S01678655(03)001661.
6. Pare, S., Bhandari, A. K., Kumara, A., & Singh, G. K. (2017). A new technique for multi-level color image thresholding based on modified fuzzy entropy and Lévy flight firefly algorithm. *Computers & Electrical Engineering, 70,* 476–495.
7. Bohat, V. K., & Arya, K. V. (2019). A new heuristic for multilevel thresholding of images. *Expert Systems with Applications, 117,* 176–203.
8. Horng, M., & Jiang, T. (2010). Multi-level image thresholding selection based on the firefly algorithm. In *Proceedings of 7th International Conference on Ubiquitous Intelligence & Computing and 7th International Conference on Autonomic & Trusted Computing*, pp. 26–29. Xian (China): IEEE Publisher, pp. 58–63.
9. Li, L., & Li, D. (2008). Fuzzy Entropy Image Segmentation based on Particle Swarm Optimization. *Progress in Natural Science, 18*(9), 1167–1171.
10. Di Martino, F., & Sessa, S. (2019). PSO image thresholding on images compressed via fuzzy transforms. *Information Sciences, 506,* 308–324.
11. Ghamisi, P., Couceiro, M. S., Martins, F. M. L., Benediktsson, J. A. (2014). Multi-level image segmentation based on fractional-order Darwinian particle swarm optimization. *IEEE Transactions on Geoscience and Remote Sensing, 52*(5), 2382–2394.
12. Rajaa, N. S. M., Sukanya, S. A., & Nikita, Y. (2015). Improved PSO based multi-level thresholding for cancer infected breast thermal images using Otsu. *Procedia Computer Science, 48*(2015), 524–529.
13. Suresh, S. S., & Lal, S. (2017). Multi-level thresholding based on chaotic Darwinian particle swarm optimization for segmentation of satellite images. *Applied Soft Computing, 55*(2017), 503–522.
14. MacQueen, J. B. (1967), Some methods for classification and analysis of multivariate observations. In *Proceedings of 5-th Berkeley Symposium on Mathematical Statistics and Probability* (vol. 1, pp. 281–297). Berkeley: University of California Press.
15. Li, H., He, H., & Wen, Y. (2015). Dynamic particle swarm optimization and K-means clustering algorithm for image segmentation. *Optik, 126*(24), 4817–4822.
16. Capor Hrosik, R., Tuba, E., Dolicanin, E., Jovanovic, R., & Tuba, M. (2019). Brain image segmentation based on firefly algorithm combined with K-means clustering. *Studies in Informatics and Control, 28*(2) 167–176. ISSN: 1220-1766.
17. Bezdek, J. C. (1981). *Pattern Recognition with Fuzzy Objective Function Algorithms*. New York: Plenum Press.
18. Ahmed, M. N., Yamany, S. M., Mohamed, N., Farag, A. A., & Moriarty, T. (2002). A modified fuzzy c-means algorithm for bias field estimation and segmentation of MRI data. *IEEE Transactions on Medical Imaging, 21,* 193–199.
19. Cai, W., Chen, S. C., & Zhang, D. Q. (2007). Fast and robust fuzzy c-means clustering algorithms incorporating local information for image segmentation. *Pattern Recognition, 40,* 825–838.

20. Chen, S. C., & Zhang, D. Q. (2004). Robust image segmentation using FCM with spatial constraints based on new kernel-induced distance measure. *IEEE Transactions on Systems, Man and Cybernetics, Part B, 34*(4), 1907–1916.
21. Szilági, L., Benyó, Z., Szilágii, S. M., & Adams, H. S. (2003). MR brain image segmentation using an enhanced Fuzzy C-Means algorithm. In *Proceedings of 25th Annual International Conference of IEEE—EMBS*, pp. 17–21.
22. Di Martino, F., Loia, V., & Sessa, S. (2010). A segmentation method for images compressed by fuzzy transforms. *Fuzzy Sets and Systems, 161*(1), 56–74.
23. Wu, K. L., & Yang, M. S. (2005). A fuzzy validity index for fuzzy clustering. *Pattern Recognition Letters, 26,* 1275–1291.

Chapter 7
Fuzzy Transforms for Image Watermarking and Image Autofocus

Digital image watermarking [6, 23] is a method applied to avoid and discover tampering in images and videos in which an authentication key is encapsulated to prevent alterations or to verify if the image was altered and localize the tampered zone. Usually the digital watermarking is classified into three categories [6, 23]:

- *robust watermarking*: it is applied to preserve the image copyright. A robust watermarking algorithm is performed in order that the encapsulated information cannot be destroyed from any computer attack;
- *fragile watermarking*: it is applied to detect the presence of alterations in the image and the encapsulated information can be easily destroyed;
- *semi-fragile watermarking*: like the fragile one, it is applied to detect the presence of alterations in the image and it is more robust with respect to user's manipulations, so that it is possible only to detect those malicious ones (innocuous manipulations as, for instance, compression operations are ignored as well).

7.1 Image Watermarking Approaches: The Fragile Block-Wise Image Watermarking

Fragile watermarking methods can be classified into the following:

- *block-wise algorithms* in which the image is partitioned into blocks and the encoding process is performed inserting a secret random signature in any block. The secret key is inserted so that the marked image appears identical to the original image to an external user;
- *pixel-wise algorithms* in which a secret binary authentication watermark is produced by pixel-value differences.

© Springer Nature Switzerland AG 2020   103
F. Di Martino and S. Sessa, *Fuzzy Transforms for Image Processing and Data Analysis*,
https://doi.org/10.1007/978-3-030-44613-0_7

The secret key is always inserted so that the marked image appears identical to the original image to an external user.

After ending the encoding process, the marked image is distributed to the users and the source image is stored to be used for successive operations of tamper detections and localizations.

Pixel-wise algorithms can localize more accurately than block-wise ones, but they are generally too expensive in terms of time execution and memory storage.

Many block-wise algorithms are proposed in the literature. The base scheme of a block-wise watermarking algorithm was proposed by Walton [1]: a pseudo-random number generated by a random walk is applied on the 2×2 blocks of the image in dependence to a secret key. The checksum is built from the seven most significant bits and is inserted in the Least Significant Bit (LSB) of the pixels.

The defect of this algorithm is the non-robustness with respect to vector quantization counterfeiting attacks. Some variation of the Wang scheme was proposed in the literature to correct this defect. For example, in [2] the authentication data of each block is generated by using a cryptographic hash function.

A well-known variation of the Walton scheme is the Chen and Wang algorithm, [3] in which the FCM algorithm is used in order to extract the authentication data.

The idea of Chen and Wang is to apply the FCM algorithm to the 2×2 blocks of the image: each block constitutes a pattern composed of four features, which are the values of the gray levels of the pixels that compose it.

Let K be the number of the blocks of the image and C be the number of clusters. After executing the FCM algorithm, Chen and Wang use the $K \times C$ final partition matrix to generate a sequence of integers $(f_1, f_2, \ldots, f_K)$ in the range $[0, 255]$, where f_k, $k = 1, \ldots, K$ is given by calculating the difference between the max and min values in the kth column of the partition matrix and multiplying this value by 255, approximating the result to the nearest integer.

Another sequence $(r_1, r_2, \ldots, r_K)$, with $r_k \in \{0, 1, \ldots, 255\}$, is created by applying a pseudo-random number generator (shortly, PRNG) seeded with a secret key (shortly, SK). Finally, an 8-bit authentication data is given by

$$d_k = f_k \oplus r_k \qquad (7.1)$$

for $k = 1, \ldots K$, where the symbol $\oplus$ is the binary XOR operator. Each two bits couple in the 8-bit authentication data d_k is embedded in the two Last Significant Bits (shortly, LSB) of the four pixels in the kth block.

Below the pseudocode of the Chen and Wang algorithm is shown.

Algorithm: Chen & Wang Block-Wise Watermarking	
Input:	*Source image I*
	Number of clusters C
	Secret Key SK
Output:	*Marked image*
1	Partition the image I in K blocks
2	Execute FCM with C clusters
3	*for* k = 1 to K
4	max = max(u_{k1},..., u_{kC})
5	min = min(u_{k1},..., u_{kC})
6	f_k := Round(255*(max-min))
7	r_k := PRNG(SK)
8	d_k := $f_k \oplus r_k$
9	Embed d_k in the four LSB's of the pixels of the k*th* block
10	*next* k
11	*return* marked I

The Chen and Wang algorithm is robust with respect to vector quantization counterfeiting attacks, but, for large images, the use of the FCM algorithm can be very expensive from a computational point of view. Some authors propose hybrid processes to optimize the cluster parameters: for example, in [4] a fragile watermarking hybrid method integrating the *K*-means clustering algorithm and Exponential Particle Swarm Optimization algorithm (EPSO) is proposed (the EPSO algorithm is used to optimize the parameters in the rough *K*-means algorithm). The main problem of these algorithms is generally the execution time which increases exponentially as the size of the image increases.

7.2 Image Tamper Detection

When a fragile image watermarking algorithm is used, to detect if the marked image is tampered, we can reconstruct the marked image and compare it with the tampered one. We obtain as a result, a binary image showing the tampered zones.

Figure 7.1 illustrates this process with an example. The original image is marked and published: subsequently, it is tampered by adding the two black and white marbles. The result is a binary image in which the tampered zones are highlighted, corresponding to those covered by the two marbles.

We cannot distinguish if the image is really tampered or if the tampered zones in the resultant binary image are caused by contrast enhancement or other simple image processing operators.

Some authors propose variations in the watermarking scheme in order to improve the detection of tampered zones. In [5, 6] a hybrid combination of a fragile and robust watermarking approach to distinguish tampers from innocent manipulations

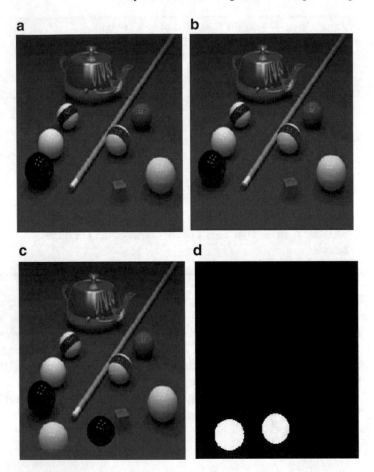

Fig. 7.1 **a** Source image, **b** Marked image, **c** Tampered image, **d** Final results

is used. In [7] a hierarchical digital watermarking method algorithm is used in order to select only tampered zones and perform the recovery of them in a percentage over 93%. In a recent work [8] propose a fragile watermarking scheme in which the maximum coefficients of the Faber-Schauder discrete wavelet transform are used to mark the image: the authors show that this scheme allows to localize correctly the image tampered zone in real-time applications. In [9] a new fragile watermarking for color images scheme is proposed to detect common image tampering operations such as copy and paste attack and image splicing attack.

Generally, a too complex watermarking scheme can increase the execution time necessary to mark the image and be not suitable to mark large size images.

In [10] a simple watermarking scheme on images compressed in the transform domain is proposed. In [11] a watermarking scheme based on table lookup in frequency domain is proposed, where the marked image is stored in compressed form.

7.3 Fuzzy Transform Method in Image Watermarking

In order to optimize the computational complexity of the watermarking process, Di Martino and Sessa propose in [12] a watermarking scheme based on the Chen and Wang algorithm in which the source image is compressed via the bidimensional direct F-transform and the compressed image is subsequently marked by applying the Chen and Wang algorithm, then the marked image is decompressed via the bidimensional inverse F-transform and distributed. In Fig. 7.2 this watermarking scheme is schematized.

The compressed marked image is stored in a compressed image dataset, so to be used in tamper detection processes. When a distributed image is tampered, it is coded via the bidimensional direct F-transform and the compressed marked image is extracted from the compressed image dataset. The two compressed images are

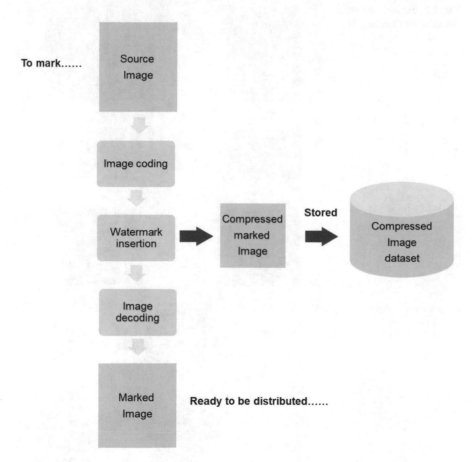

Fig. 7.2 Watermarking scheme in [12]

compared with the tamper detection. Finally, the tamper localization function produces the binary tamper localization image in which the tampered zones are shown. In Fig. 7.3 this tamper detection process is schematized.

To obtain the better compression rate, the following preprocessing strategy is applied.

Let $(PSNR)_0$ be the PSNR calculated by measuring the quality of the marked image with respect to the source image without compression of the original image (compression rate $\rho = 1$). Afterwards the image is compressed more strongly at each iteration. In the cycle, the compressed image is marked and subsequently decompressed to calculate the PSNR obtained by comparing it with the source image.

The compression is considered acceptable if the RMS obtained in the calculation of the quality of the image is less or equal to 2.5 times the value of the RMSE

Fig. 7.3 Scheme of tamper detection process in [12]

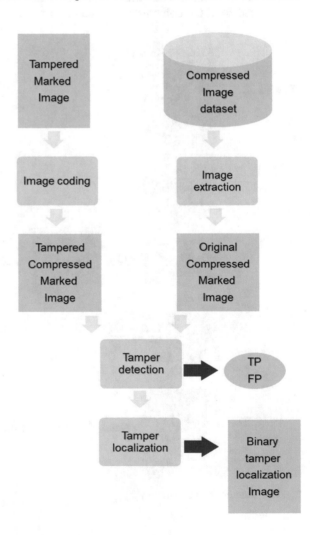

obtained by marking directly the source image, that is RMSE is less or equal to the value $(RMSE)_1 = 2.5 \cdot (RMSE)_0$, corresponding to the PSNR value given by

$$(PSNR)_1 = \frac{1}{\log_{10}\left(\frac{1}{2.5}\right)} \left(\frac{(PSNR)_0}{1 - (PSNR)_0} \right) \tag{7.2}$$

If during an iteration we obtain PSNR < $(PSNR)_1$, then the process ends and the optimal compression rate is given by the compression rate used in the previous iteration. Below in the pseudocode, the watermarking algorithm proposed in [12] is schematized. Input of the algorithm is the source image I_0, outputs are the best compression rate ρ_{BEST} and the corresponding marked compressed image I_M.

	Algorithm: Di Martino & Sessa Block-Wise Insertion mark	
Input:		$N \times M$ source image I_0
Output:		Best compression rate ρ_{BEST}
		Final compressed marked image I_M
1	Mark the source image I_0 via the Chen & Wang algorithm	
2	$I_M :=$ compressed marked image	
3	Calculate PSNR = $(PSNR)_0$	
4	Calculate $(PSNR)_1$ via (7.2)	
5	$\rho_{BEST} := 1$	
6	Use an initial compression rate $\rho < 1$	
7	*while* PSNR $\geq (PSNR)_1$	
8	Compress the source image I_0 via direct F-transform	
9	Mark the compressed image via the Chen & Wang algorithm	
10	Decompress the compressed image via inverse F-transform	
11	Calculate PSNR	
12	*if* PSNR $\geq (PSNR)_1$ *then*	
13	$\rho_{BEST} := \rho$	
14	$I_M :=$ compressed marked image	
15	Decrease ρ	
16	*end if*	
17	*end while*	
18	*return* ρ_{BEST}, I_M	

The final compressed image I_M is stored and subsequently decompressed via the inverse F-transform and ready to be published.

Tamper detection activities are carried out by decompressing the compressed marked image I_M and extracting the zones tampered in the corresponding tampered image; the results of this process are the binary image with the tampered zones localized.

To measure the performance of the tamper detection process [12], a set of experiments are performed in which this binary image is compared with the one in which the tampered zone is exactly localized. The sensitivity and specificity indexes are calculated with the following:

$$\text{sensitivity} = \frac{TP}{TP + FN} \tag{7.3}$$

$$\text{specificity} = \frac{TN}{TN + FP} \tag{7.4}$$

They are calculated measuring the True Positive (*TP*), False Positive (*FP*), True Negative (*TN*), and False Negative (*FN*) rates in the comparison between the two resulting binary images. High values of sensitivity and specificity imply that the information loss due to compression does not affect the correct detection and localization of the tampered zones.

In Fig. 7.4a the 512×512 color image Baboon is shown. Figure. 7.4b–d shows the source image, respectively, in the three bands R, G, and B.

In Fig. 7.5a the tampered image is shown. Figure 7.5b–d are the tampered image in the R, G, B bands, respectively.

In Fig. 7.6a–c the tampered zones in each band obtained by marking directly the

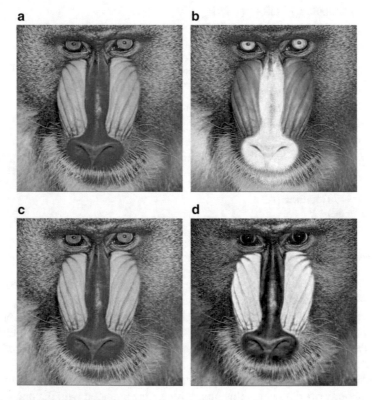

Fig. 7.4 **a** Source image Baboon, **b** R band, **c** G band, **d** B band

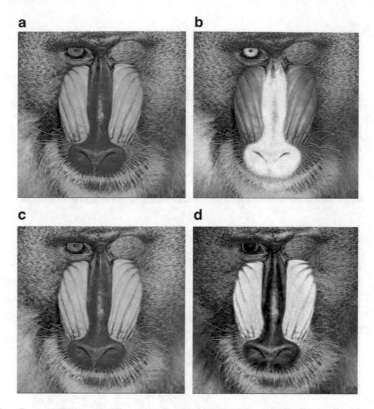

Fig. 7.5 **a** Tampered image, **b** R band (tampered), **c** G band (tampered), **d** B band (tampered)

source image are shown without compression.

Finally, in Fig. 7.7a–c the tampered zones in each band obtained by marking directly the image compressed with a compression rate $\rho = 0.25$ are shown and ρ the best compression rate is obtained.

Table 7.1 shows the mean values of the *TP*, *FP*, *TN*, *FN* rates and the mean sensitivity and specificity indexes are calculated by varying the compression rates, where the mean is obtained considering the indexes measured in each band.

The sensitivity and specificity obtained without compression of the original image ($\rho = 1$) are 99.4% and 100%, respectively, those obtained with the optimal compression rate $\rho = 0.25$ are 99.9% and 95%, respectively.

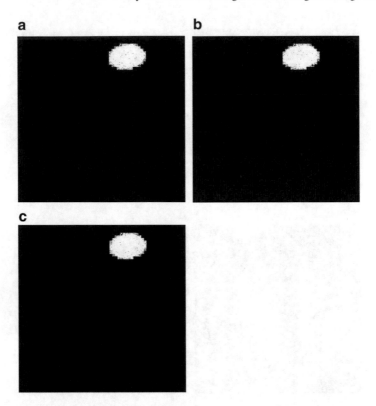

Fig. 7.6 **a** No compression, *R* band Fig. 7.6, **b** No compression, *G* band, **c** No compression, *B* band

7.4 Fuzzy Transform Image Watermarking via Fuzzy Bilinear Equations

In order to improve the performances of the F-transform watermarking algorithm, in ([13] a new *F*-transform watermarking algorithm) is proposed in which the maximum solution of the Bilinear Fuzzy Relation Equations (BFRE) is applied to any 2×2 block in the image compressed via the direct *F*-transform in order to mark its pixels.

The BFRE algorithm [14] finds the maximum solution of a system of n bilinear fuzzy relation equations with n unknowns.

Let *A*, *B* be two $n \times n$ fuzzy matrices, where $A = [a_{ij}]$ and $B = [b_{ij}]$, a_{ij}, $b_{ij} \in$ [0, 1] and $i, j = 1, 2, ..., n$. Let $x = (x_1, x_2, ..., x_n)$ be a vector of n unknowns. The following system of fuzzy relation equations:

$$a_i \vee \bigcup_{j=1}^{n} (a_{ij} \wedge x_j) = b_i \vee \bigcup_{j=1}^{n} (b_{ij} \wedge x_j) \tag{7.5}$$

for $i = 1, ..., m$, is called *system of external fuzzy bilinear equations.*

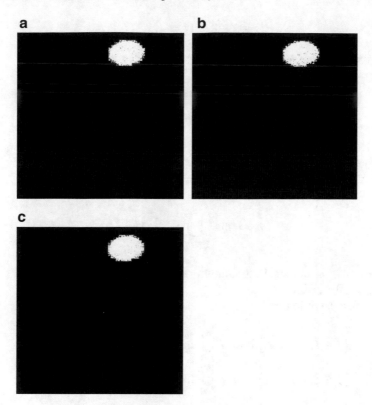

Fig. 7.7 **a** $\rho = 0.25$, R band, **b** $\rho = 0.25$, G band, **c** $\rho = 0.25$, B band

Table 7.1 Tamper detection results for the color image Baboon

ρ	TP	FP	TN	FN	Sensitivity	Specificity
1.000	99.392	0.041	99.959	0.608	0.994	1.000
0.250	95.045	0.067	99.929	4.955	0.950	0.999
0.063	94.525	0.071	99.933	5.475	0.945	0.999
0.016	97.949	0.473	99.770	6.175	0.941	0.995

The value

$$\rho_i = \min\left(a_i \vee \bigcup_{j=1}^{n} a_{ij}, \; b_i \vee \bigcup_{j=1}^{n} b_{ij} \right) \tag{7.6}$$

for $i = 1, \ldots, m$, is called the *critical value of the ith fuzzy bilinear equation* in (7.5). We call

$$\rho_k = \min\{\rho_i : i = 1, \ldots, m\} \tag{7.7}$$

The *critical value related to the kth equation* in the system (7.5).
If $a_i = b_i = 0$ for each $i = 1, \ldots, m$, (7.5) becomes

$$\bigcup_{j=1}^{n} a_{ij} \wedge x_j = \bigcup_{j=1}^{n} b_{ij} \wedge x_j \tag{7.8}$$

called *system of fuzzy bilinear equations*.
   The critical value of the *i*th fuzzy bilinear equation Eq. (7.8) is given by

$$\rho_i = \min\left(\bigcup_{j=1}^{n} a_{ij}, \bigcup_{j=1}^{n} b_{ij}\right) \tag{7.9}$$

for $i = 1, \ldots, m$, the smallest solution of the system (7.8) is the trivial solution
$\mathbf{0} = (0, \ldots, 0) \in [0, 1]^n$.
   We define the following:

$$\Delta_i = \begin{cases} \left\{j \in I_n | b_{ij} > r_i\right\} \text{ if } a_i \vee \bigcup_{j=1}^{n} a_{ij} \leq b_i \vee \bigcup_{j=1}^{n} b_{ij} \\ \left\{j \in I_n | a_{ij} > r_i\right\} \text{ if } a_i \vee \bigcup_{j=1}^{n} a_{ij} \geq b_i \vee \bigcup_{j=1}^{n} b_{ij} \end{cases} \quad i = 1, 2, \ldots, m \tag{7.10}$$

called *the difference index set of critical value* of the *i*th fuzzy bilinear equation in
(7.5). For finding the greatest solution $\tilde{x} = (\tilde{x}_1, \ldots, \tilde{x}_n)$ of the system (7.5), in [14]
the following algorithm is applied:

Algorithm: BFRE (Li, 1992)	
Input:	*A system of external fuzzy bilinear equations.*
Output:	*The maximum solution* $\tilde{\tilde{x}}$
1	unknown := n // number of unknown variables in (7.5)
2	**while** unknown > 0
3	Calculate ρ_k with (15)
4	Calculate the corresponding set $\Delta_k = \{j_1, j_2, \ldots, j_t\}$.
5	**for each** j in Δ_k
6	$\tilde{x}_i := \rho_k$
7	Substitute $\tilde{x}_j$ with ρ_k in the system (7.5)
8	unknown := unknown -1
9	**next**
10	**end while**
11	**return** $\tilde{x} = (\tilde{x}_1, \ldots, \tilde{x}_n)$

Di Martino and Sessa [13] apply the BFRE algorithm in a fragile F-transform based block-wise watermarking method for color images. The source image is compressed in each band by the bidimensional direct F-transform, then the BFRE algorithm is applied. The compressed images in the bands R, G, B are partitioned in blocks 2×2. The authentication data of the kth block in the compressed R band is found by applying the BFRE algorithm to the system (7.8) in which the a_i, $i = 1, \ldots,$ 4, are the normalized values of the pixels in this block and the b_i, $i = 1, \ldots, 4$, are the normalized values of the pixels in the corresponding kth block in the compressed G band.

The method to find the authentication data is schematized in Fig. 7.8. After applying the BFRE algorithm to the bilinear fuzzy system with $n = 2$ in the form (7.8), the mean of the components of the maximum solution is calculated. Then it is de-normalized, obtaining an integer in $\{0, 1, \ldots, 255\}$ used to obtain the final 8-bit authentication data. The authentication data is subsequently used to mark the kth block of the compressed R band.

In practice, if each compressed image is decomposed in $2 \times 2 \, K$ blocks, the sequence of K integers obtained in [12] by applying the Chen and Wang algorithm is now obtained in each band using the BFRE algorithm applied to 2×2 fuzzy bilinear equations.

The watermark insertion process is schematized in Fig. 7.9.

Below in the pseudocode, the BFR insertion mark algorithm is shown, where we consider the G band as next band to R, the B band as next band to G, and the R band as next band to B.

Algorithm: BFRE insertion mark	
Input:	$N{\times}M$ source color image
Output:	Final marked color image
1	The original image is compressed in any band via direct F-Transform
2	The compressed bands are partitioned in K blocks 2×2
3	*for* each band
4	*for* each block
5	Create the BRFE system (7.8) using the k*th* block in next band
6	Execute the BFRE algorithm
7	Calculate the mean value $\tilde{x}$ of the max solution components
8	Calculate s de-normalizing $\tilde{x}$
9	Calculate the authentication data
10	Mark the block
11	*next*
12	Decompress the compressed marked band via inverse F-transform
13	*next*
14	*return* the marked color image

The same strategy used in the block-wise insertion mark algorithm in [12] is used to find the best compression rate. The PSNR threshold $(PSNR)_1$ is calculated via (7.2).

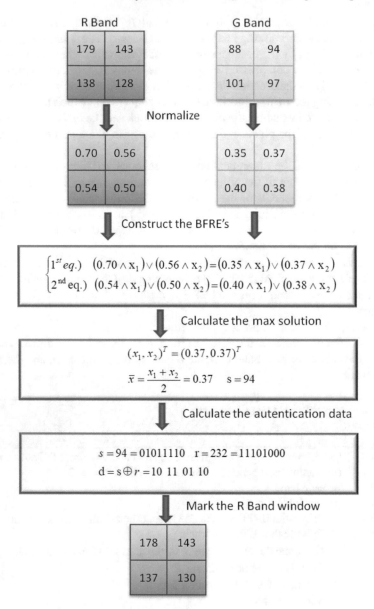

Fig. 7.8 Example of strategy for finding the authentication data in [13]

The best compression rate is given by the smallest compression rate for which in all three bands the PSNR value is greater or equal to $(PSNR)_1$.

The tamper detection and localization processes are performed compressing the tampered image, extracting, and marking the compressed images from the image dataset and detecting the tampered zones comparing the two compressed images in

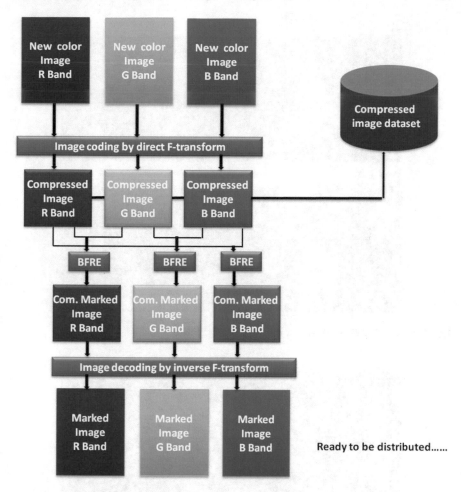

Fig. 7.9 Watermark insertion scheme in [13]

any band. A set of comparison tests between the BFRE and F-transform watermarking algorithms are carried out in [13].

Below the results are shown applying the BFRE insertion mark and tamper detection algorithms to the 800×600 color image Rose of Fig. 7.10a.

The best compression rate obtained in the preprocessing phase is $\rho = 0.25$. In Fig. 7.11a the marked image is shown. Figure. 7.11b–d gives the marked image in each band.

In Fig. 7.12a the tampered image is shown; Fig. 7.12b–d shows the tampered image in each band.

Figure 7.13a–c shows the results of the tamper detection and localization processes in each band.

Fig. 7.10   **a** Source image Rose, **b** *R* band, **c** *G* band, **d** *B* band

Fig. 7.11   **a** Marked image Rose, **b** *R* band, **c** *G* band, **d** *B* band

Fig. 7.12 **a** Tampered image Rose, **b** *R* band, **c** *G* band, **d** *B* band

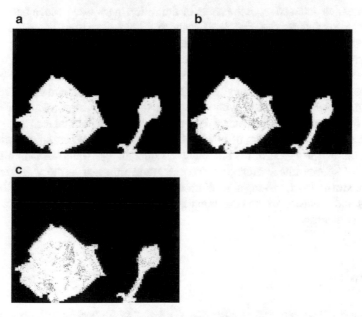

Fig. 7.13 **a** Tampered zone—*R* band, **b** Tampered zone—*G* band a Tampered zone—*B* band

Table 7.2 Comparison results for the color image Rose

ρ	BFRE [13]	F-transform [12]				
	Mean sensitivity (%)	Mean specificity (%)	Mean accuracy (%)	Mean sensitivity (%)	Mean specificity (%)	Mean accuracy (%)
1.00	99.31	99.96	99.64	99.16	99.95	99.56
0.25	95.46	99.93	97.70	94.27	99.91	97.09
0.16	95.13	99.92	97.53	93.88	99.90	96.89
0.063	94.62	99.91	97.27	93.21	99.88	96.55
0.016	92.01	99.59	95.80	90.18	99.03	94.61

The comparison test results obtained in [13] for the color image Rose are shown in Table 7.2. This table contains the mean sensitivity, specificity, and accuracy obtained for the compression rates where the sensitivity and specificity indexes are in (7.3) and (7.4) and the accuracy index is given by

$$accuracy = \frac{TP + TN}{TN + TP + FP + FN} \qquad (7.11)$$

The mean is deduced from the mean of the index measured in each band.

These results show that the BFRE watermarking algorithm has better performances than the F-transform watermarking algorithm. In fact, the average values of all three indicators obtained with the first method are higher than those obtained with the second method. Furthermore, for all three indicators the difference between the average values obtained using the BFRE method compared and that used by the F-transform method increases with increasing the compression rate. This trend is verified for all images of the dataset used during the experimental test.

Since the BFRE algorithm of block-wise fragile image watermarking preserves the benefits of the F-transform block-wise fragile image watermarking algorithm in terms of source image archiving and execution times, it is more efficient than the F-transform watermarking algorithm and represents an optimal tradeoff between accuracy and precision in the detection of tampered zones and execution times and memory allocation.

References

1. Walton, S. (1995). Information authentication for a slippery new age. *Dr. Dobbs Journal, 20*(4), 18–26.
2. Chang, C. C., Hu, Y. S., & Lu, T. C. (2006). A watermarking-based image ownership and tampering authentication scheme. *Pattern Recognition Letters, 27*(5), 439–446.
3. Chen, W. C., & Wang, M. S. (2009). A fuzzy c-means clustering-based fragile watermarking scheme for image authentication. *Expert Systems with Applications, 36,* 1300–1307.

4. El Bakrawy, L.M., Ghali, N.I., Kim, T.-H., Hassanien, A.E. (2011), A block-wise-based fragile watermarking hybrid approach using rough sets and exponential particle swarm optimization. *International Journal of Future Generation Communication and Networking, 4*(4), 77–88.
5. Fridrich, J. (1999). A hybrid watermark for tamper detection in digital images. *Fifth International Symposium on Signal Processing and its Applications (ISSPA)*. https://doi.org/10.1109/ISSPA.1999.818172.
6. Deguillaume, F., Voloshynovskiy, S., & Pun, T. (2003). Secure hybrid robust watermarking resistant against tampering and copy attack. *Signal Processing, Elsevier, 83*(2003), 2133–2170.
7. Phen, L. L., Hsiehb, C.-K., & Huangb, P.-W. (2005). A hierarchical digital watermarking method for image tampers detection and recovery. *Pattern Recognition, 38*(12), 2519–2528.
8. Azeroual, A., & Afdel, K. (2017). Real-time image tamper localization based on fragile watermarking and Faber-Schauder wavelet. *AEU—International Journal of Electronics and Communications, 79,* 207–218.
9. Manikandan, V.M., & Masilamani, V. (2016). A context dependent fragile watermarking scheme for tamper detection from demosaicked color images. In *ICVGIP '16 Proceedings of the Tenth Indian Conference on Computer Vision, Graphics and Image Processing*. Article No. 61. https://doi.org/10.1145/3009977.3009987.
10. Wolfgang, R.B., & Delp, E.J. (1996). A watermark for digital images. In *Proceedings of IEEE International Conferences on Image Processing* (Vol. III, pp. 219–222). IEEE Press.
11. Wong, P. W., & Memon, N. (2001). Secret and public key image watermarking schemes for image authentication and ownership verification. *IEEE Transactions on Image Processing, 10*(10), 1593–1601.
12. Di Martino, F., & Sessa, S. (2012). Fragile watermarking tamper detection with images compressed by fuzzy transform. *Information Sciences, 195,* 62–90.
13. Di Martino, F., & Sessa, S. (2019). Fragile watermarking tamper detection via bilinear fuzzy relation equations. *Journal of Ambient Intelligence and Humanized Computing, 10*(5), 2041–2061.
14. Li, J. X. (1992). A new algorithm for the greatest solution of fuzzy bilinear equations. *Fuzzy Sets and Systems, 46,* 193–210.

Chapter 8
Fuzzy Transform for Data Analysis

This chapter focuses on the use of the F-transform in data analysis. In previous chapters, we have seen how the discrete bidimensional F-transform can be used in many image analysis applications.

We know that a direct mono-dimensional F-transform converts an assigned continuous function, defined in the interval $[a, b]$ and known in a discrete set of N points, in a convenient n-dimensional vector and the inverse F-transform converts this n-dimensional vector into another continuous function which approximates the original function. Extending this concept to the bidimensional case, we have applied the direct and inverse discrete bidimensional F-transforms for coding and decoding images, applying them in various image analysis activities as image compression, image fusion, image watermarking, and image focusing.

Now we show that we can apply these concepts in data analysis extending the F-transform in a multi-dimensional space.

8.1 Multi-dimensional Fuzzy Transform Applied in Data Analysis

Formaking this self-contained, we repeat here some concepts and definitions. Let $f(x_1, x_2, \ldots, x_k)$ be a function of $k > 2$ variables defined in the Cartesian product $[a_1, b_1] \times [a_2, b_2] \times \cdots \times [a_k, b_k]$ of R^K.

Let $X_{11}, X_{22}, \ldots, X_{1n1} \in [a_1, b_1], \ldots, X_{k1}, X_{k2}, \ldots, x_{kn_k} \in [a_k, b_k]$ be $n_1 + \ldots + n_k$ assigned points, called nodes, such that $X_{i1} = a_i < X_{i2} < \ldots < x_{in_i} = b_i$ for $i = 1, \ldots, k$. Furthermore, let $\{A_{i1}, A_{i2}, \ldots, A_{in_i}\}$ be a fuzzy partition of $[a_i, b_i]$ for $i = 1, \ldots, k$.

We call $(h_1, h_2, \ldots, h_k)$th component $F_{h_1, h_2, \ldots, h_K}$ of the direct F-transform of f with respect to the basic functions $\{A_{11}, A_{12}, \ldots, A_{1n_1}\}, \ldots, \{A_{k1}, A_{k2}, \ldots, A_{kn_k}\}$ to be the following value:

© Springer Nature Switzerland AG 2020

F. Di Martino and S. Sessa, *Fuzzy Transforms for Image Processing and Data Analysis*,
https://doi.org/10.1007/978-3-030-44613-0_8

$$F_{h_1 h_2 \dots h_k} = \frac{\int_{a_1}^{b_1} \int_{a_2}^{b_2} \dots \int_{a_k}^{b_k} f(x_1, x_2, \dots, x_k) A_{1_{h_1}}(x_1) A_{2_{h_2}}(x_2) \dots A_{k_{h_k}}(x_k) dx_1 dx_2 \dots dx_k}{\int_{a_1}^{b_1} \int_{a_2}^{b_2} \dots \int_{a_k}^{b_{kk}} A_{1_{h_1}}(x_1) A_{2_{h_2}}(x_2) \dots A_{k_{h_k}}(x_k) dx_1 dx_2 \dots dx_k} \qquad (8.1)$$

Furthermore, we call *the multi-dimensional inverse F-transform* of f with respect to the basic functions $\{A_{11}, A_{12}, \dots, A_{1n_1}\}, \dots, \{A_{k1}, A_{k2}, \dots, A_{kn_k}\}$ the following function defined for every $(x_1, x_2, \dots, x_k) \in [a_1, b_1] \times [a_2, b_2] \times \dots \times [a_k, b_k]$:

$$f_{n_1 n_2 \dots n_K}^F (x_1, x_2, \dots, x_k) = \sum_{h_1=1}^{n_1} \sum_{h_2=1}^{n_2} \sum_{h_k=1}^{n_k} F_{h_1, h_2, \dots, h_K} \cdot A_{1_{h_1}}(x_1) \cdot \dots \cdot A_{k_{h_k}}(x_k) \qquad (8.2)$$

Similarly to the one-dimensional case, the *multi-dimensional inverse F-transform* of f, can approximate the function f.

In the discrete case, the function f is known in m points $P = \{p_1, p_2, \dots, p_m\}$ with $p_j = (p_{j1}, p_{j2}, \dots, p_{jk}) \in [a_1, b_1] \times [a_2, b_2] \times \dots \times [a_k, b_k]$. We say that the set P is sufficiently dense with respect to the fuzzy partitions $\{A_{11}, A_{12}, \dots, A_{1n_1}\}$, $\dots, \{A_{k1}, A_{k2}, \dots, A_{kn_k}\}$ if for each k-tuple $\{h_1, \dots, h_k\} \in \{1, \dots, n_1\} \times \dots \times \{1, \dots, n_k\}$ there exists a point $p_j = (p_{j1}, p_{j2}, \dots, p_{jk}) \in P, j \in \{1, \dots, m\}$, such that $A_{1h_1}(p_{j1}) \cdot A_{2h_2}(p_{j2}) \cdot \dots \cdot A_{kh_k}(p_{jk}) > 0$.

In this case we can define the *discrete direct F-transform* of f with respect to the fuzzy partitions $\{A_{11}, A_{12}, \dots, A_{1n_1}\}, \dots, \{A_{k1}, A_{k2}, \dots, A_{kn_k}\}$. The $(h_1, h_2, \dots, h_k)$th component $F_{h_1 \dots h_k}$ of the discrete direct F-transform of f is given by

$$F_{h_1 h_2 \dots h_K} = \frac{\sum_{j=1}^{m} f(p_{j1}, p_{j2}, \dots p_{jk}) \cdot A_{1h_1}(p_{j1}) \cdot A_{2h_2}(p_{j2}) \cdot \dots \cdot A_{kh_K}(p_{jk})}{\sum_{j=1}^{m} A_{1h_1}(p_{j1}) \cdot A_{2h_2}(p_{j2}) \cdot \dots \cdot A_{kh_K}(p_{jk})}$$

$$(8.3)$$

We call *discrete inverse F-transform of f* with respect to the basic functions $\{A_{11}, A_{12}, \dots, A_{1n_1}\}, \dots, \{A_{k1}, A_{k2}, \dots, A_{kn_K}\}$ the following function by setting for each point $p_j = (p_{j1}, p_{j2}, \dots, p_{jk}) \in [a_1, b_1] \times \dots \times [a_k, b_k]$ with $j = 1, 2, \dots m$:

$$f_{n_1 n_2 \dots n_K}^F (p_{j1}, p_{j2}, \dots, p_{jk}) = \sum_{h_1=1}^{n_1} \sum_{h_2=1}^{n_2} \dots \sum_{h_K=1}^{n_k} F_{h_1 h_2 \dots h_K} \cdot A_{1h_1}(p_{j1}) \cdot \dots \cdot A_{kh_K}(p_{jk}) \qquad (8.4)$$

The following theorem holds:

Theorem 8.1 Let $f(x_1, x_2, \dots, x_k)$ be a function assigned on the set of points $P = \{(p_{11}, p_{12}, \dots, p_{1k}), (p_{21}, p_{22}, \dots, p_{2k}), \dots, (p_{m1}, p_{m2}, \dots, p_{mk})\}$ of the Cartesian product $[a_1, b_1] \times [a_2, b_2] \times \dots \times [a_k, b_k]$. Then for every $\varepsilon > 0$, there exist k integers $n_1(\varepsilon), \dots, n_k(\varepsilon)$ and related fuzzy partitions $\{A_{11}, A_{12}, \dots, A_{1n_1(\varepsilon)}\}, \dots, \{A_{k1}, A_{k2}, \dots, A_{kn_k(\varepsilon)}\}$ such that the set P is sufficiently dense with respect to these fuzzy partitions and for every $p_j = (p_{j1}, p_{j2}, \dots, p_{jk}) \in P, j = 1, \dots, m$, the following inequality holds:

$$\left| f(p_{j1}, p_{j2}, \dots, p_{jk}) - f_{n_1(\varepsilon) n_2(\varepsilon) \dots n_k(\varepsilon)}^F (p_{j1}, p_{j2}, \dots, p_{jk}) \right| < \varepsilon \qquad (8.5)$$

8.2 The Inverse Multi-dimensional Fuzzy Transform for Assessing Functional Dependencies in the Data

In [1, 2] the multi-dimensional discrete F-transform is applied to find dependency between attributes in the data.

Following [1, 2], we schematize a dataset with r numerical attributes $X_1, \ldots, X_i, \ldots, X_r$ and m instances $O_1, \ldots, O_j, \ldots, O_m$, with $m > r$, in the following relation:

	X_1	$\cdots$	X_i	$\cdots$	X_r
O_1	p_{11}	.	p_{1i}	.	p_{1r}
.	.	.	.	.	.
.	.	.	.	.	.
.	.	.	.	.	.
O_j	p_{j1}	.	p_{ji}	.	p_{jr}
.	.	.	.	.	.
.	.	.	.	.	.
.	.	.	.	.	.
O_m	p_{m1}	.	p_{mi}	.	p_{mr}

We assume that the known domain of the attribute X_i, $i = 1, \ldots, n$, is the real interval $[a_i, b_i]$ defined by setting $a_i = \min\{p_{1i}, \ldots, p_{mi}\}$ and $b_i = \max\{p_{1i}, \ldots, p_{mi}\}$.

Now we suppose to analyze if exists a functional dependency of the attribute X_z, $z \in \{k + 1, \ldots, r\}$ from the set of attributes $X_1, \ldots, X_k$ $(1 \leq k \leq r)$. This functional dependency is given in the following form:

$$X_Z = H(X_1, \ldots, X_k) \tag{8.6}$$

where $H: [a_1, b_1] \times [a_2, b_2] \times \cdots \times [a_k, b_k] \rightarrow [a_z, b_z]$ is a continuous function of k variables.

In [1] the multi-dimensional inverse F-transform was applied in regression analysis to assess the functional dependency (8.6). We suppose that the function $H(X_1, \ldots, X_k)$ is known in m points $(p_{j1}, p_{j2}, \ldots, p_{jk}), j = 1, \ldots, m$, by setting $H(p_{j1}, p_{j2}, \ldots, p_{jk}) = p_{jz}$ for $j = 1, 2, \ldots, m$.

For any interval $[a_i, b_i]$, $i = 1, \ldots, k$, a fuzzy partition $\{A_{i1}, A_{i2}, \ldots, A_{in_i}\}$ is created with $n_i \geq 3$. If the set of m points is sufficiently dense with respect to these fuzzy partitions, we can define the multi-dimensional direct F-transform of H with $(h_1, h_2, \ldots, h_k)$th component given by

$$F_{h_1 h_2 \ldots h_k} = \frac{\sum_{j=1}^{m} p_{jz} \cdot A_{1_{h_1}}(p_{j1}) \cdot \ldots \cdot A_{k_{h_K}}(p_{jk})}{\sum_{j=1}^{m} A_{1_{h_1}}(p_{j1}) \cdot \ldots \cdot A_{k_{h_K}}(p_{jk})} \tag{8.7}$$

By (8.4), the inverse F-transform $H^F_{n_1 n_2 \dots n_k}$ of H is given by

$$H^F_{n_1 n_2 \dots n_k}(p_{j1}, p_{j2}, \cdots p_{jk}) = \sum_{h_1=1}^{n_1} \sum_{h_2=1}^{n_2} \cdots \sum_{h_k=1}^{n_k} F_{h_1 h_2 \dots h_K} \cdot A_{1h_1}(p_{j1}) \cdot \dots \cdot A_{kh_K}(p_{jk}) \qquad (8.8)$$

To evaluate the existence of this functional dependency in [1] the statistical index of determinacy is measured and given by

$$r_c^2 = \frac{\sum_{j=1}^{m} \left(H^F_{n_1 n_2 \dots n_k}(p_{j1}, p_{j2}, \cdots p_{jk}) - \hat{p}_z \right)^2}{\sum_{j=1}^{m} \left(p_{jz} - \hat{p}_z \right)^2} \qquad (8.9)$$

where $\hat{p}_z$ is the mean of the instance values $p_{1z}, p_{2z}, \dots, p_{mz}$ of the attribute X_z. The index of determinacy r_c^2 ranges in the interval [0, 1], where $r_c^2 = 0$ means that $H^F_{n_1 n_2 \dots n_k}$ does not fit to the data and, conversely, $r_c^2 = 1$ means that $H^F_{n_1 n_2 \dots n_k}$ fits perfectly to the data.

A variation of the formula of the index (8.9) used in multiple regression analysis [3] is given by

$$r_c'^2 = 1 - \left[\left(1 - r_c^2 \right) \cdot \frac{m-1}{m-k-1} \right] \qquad (8.10)$$

This formula includes both the number of independent variables k and the scale of the data sample.

8.3 The Problem of the Sufficient Density of Data Points with Respect to the Fuzzy Partition

While in image analysis, the data are always sufficiently dense with respect to the set of the two fuzzy partitions because the pixels are arranged in a regular manner, in data analysis it is not possible to guarantee the sufficient density of the data with respect to the fuzzy partition.

So the set of fuzzy partitions could be too fine with respect to the dataset. Conversely, fuzzy partitions that ensure compliance with sufficient data density could be too coarse grained and result in poor performance.

Figure 8.1 gives an example of data not sufficiently dense with respect to the fuzzy partition. Since no data point is included in the gray area, we have that $\sum_{j=1}^{m} A_{1h_1}(p_{j1}) \cdot A_{2h_2}(p_{j2}) = 0$.

Figure 8.2 shows an example of fuzzy partitions too coarse grained with respect to the data. Both two domains are partitioned by using 3 nodes. As we can see in the figure, it is possible to obtain finer partitions by increasing the number of nodes

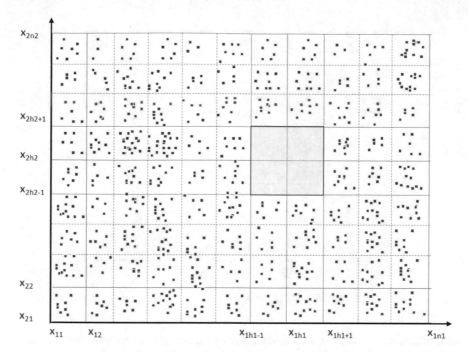

Fig. 8.1 Example of data not sufficiently dense with respect to the fuzzy partitions

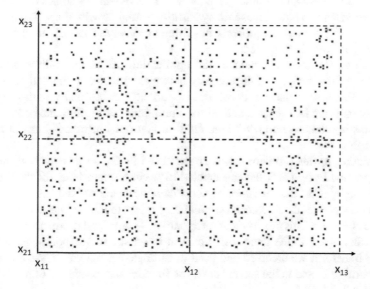

Fig. 8.2 Example of fuzzy partitions too coarse grained

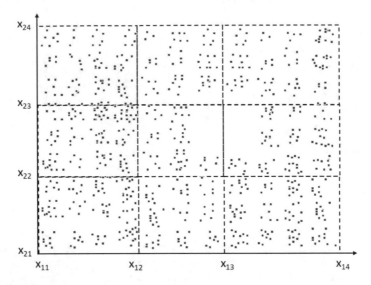

Fig. 8.3 Example of acceptable fuzzy partitions

(for example, 4 nodes for each domain) and respecting the constraint of sufficient density of data density with respect to the two fuzzy partitions.

Figure 8.3 shows an example of fuzzy partitions in which 4 nodes are used . As we can see from this figure, these fuzzy partitions are finer with respect to the fuzzy partitions in Fig. 8.2 and the property of sufficient density of the data with respect to the fuzzy partitions is respected.

Therefore, the main problem in the application of the multi-dimensional F-transform for the analysis of dependency between attributes in the data is to build fuzzy partitions such that the constraint of sufficient data density with respect to the fuzzy partition is respected and, at the same time, the performance obtained by approximating the function f with the multi-dimensional inverse F-transform (8.8) is acceptable.

A solution to this problem has been proposed in [2]. In this method, initially coarse-grained fuzzy partitions are created by setting a threshold for the index of determinacy r_c^2: the fuzzy partition is considered acceptable if the calculated index is equal to or greater than a threshold α and if r_c^2 is less to this threshold a finer fuzzy partition is created. This process is iterated until $r_c^2 \geq \alpha$ or the data are not sufficiently dense with respect to the fuzzy partitions: in the first case the components of the direct F-transform are archived and used to estimate the function f via the inverse F-transform (8.8) and in the second case, the functional dependency of X_z from X_1, ..., X_k is not detected.

This method is described in detail in the next paragraph.

8.4 Fuzzy Transform Method for the Analysis of Numerical Attribute Dependencies in Datasets

In [2] a method is proposed to detect dependency between attributes in a dataset by using the multi-dimensional F-transform. This method is based on an iterative process implemented to determine the optimal fuzzy partition of domains $[a_1, b_1]$, $[a_2, b_2]$, ..., $[a_k, b_k]$.

To reduce the computational complexity of the algorithm, the authors propose that the number of nodes of each domain is identical, namely, that $n_1 = n_2 = \cdots = n_k = n$.

Initially $n = 3$ is set, building the most coarse-grained fuzzy partitions of the k domains. Then the algorithm verifies that the data are sufficiently dense with respect to the fuzzy partitions: this control is performed by checking that for no combination $\left(A_{1h_1} A_{2h_2} \ldots A_{k_{h_K}}\right)$, $h_i = 1, \ldots, n$, holds $\sum_{j=1}^{m} A_{1h_1}\left(p_{j1}\right) \cdot A_{2h_2}\left(p_{j2}\right) \cdot \ldots \cdot A_{k_{h_K}}\left(p_{jk}\right) = 0$.

If this condition is not respected, then the data are not sufficiently dense with respect to the fuzzy partitions, otherwise the component of the direct fuzzy transform (8.7) is calculated and the index of determinacy r_c^2 or its variation $r_c'^2$ are taken into consideration. If the index of determinacy is greater or equal to a threshold α, the dependency of X_z from $X_1, \ldots, X_k$ is found and the direct fuzzy transform components are stored, otherwise the number of nodes is increased by 1, the new fuzzy partitions are constructed and the process is repeated.

In Fig. 8.4 the flow diagram of this process is shown.

Below in the pseudocode, the algorithm in [2] is shown.

The algorithm *F-transform Attribute Dependency* calls *the function Sufficient Density* $(X_1, X_2, \ldots, X_k, X_z, n)$ in order to know if the data are sufficiently dense with respect to the fuzzy partitions. The argument of this function is the number n of nodes and the subset with the attributes $X_1, X_2, \ldots, X_k$, and X_z. The function returns TRUE if the data are sufficiently dense with respect to the fuzzy partitions, otherwise it returns the value FALSE.

Fig. 8.4 Flow diagram of
the method proposed in [2]

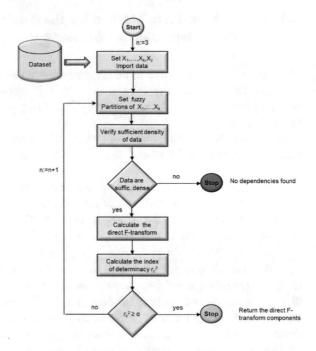

Algorithm: F-transformAttributeDependency	
Input:	*Subset of the input attributes X_1, X_2,…,X_k*
	Subset of the output attribute X_z
	Threshold value α
Output:	*Direct F-transform components*

1	n := 3
2	r_c^2 := 0
3	dependencyFound := *FALSE*
4	*while* r_c^2 < α
5	Create the fuzzy partitions of $[a_1,b_1]$, $[a_2,b_2]$,…,$[a_k,b_k]$ with n nodes
6	*if* SufficientDensity(X_1, X_2,…,X_k, X_z,n) = TRUE *then*
7	*for each* combination $(h_1..h_k)$
8	Calculate $F_{h_1 h_2 … h_k}$ via (8.7)
9	*next* $(h_1..h_k)$
10	Compute the index of determinacy via (8.9) or (8.10)
11	*if* (Index of determinacy ≥ α) *then*
12	*for each* combination $(h_1..h_k)$
13	Store $F_{h_1 h_2 … h_k}$
14	*next* $(h_1..h_k)$
15	dependencyFound := *TRUE*
16	*else*
17	n := n+1
18	*end if*
19	*else*
20	dependencyFound := *FALSE*
21	*end if*
22	*return* dependencyFound

Algorithm: SufficientDensity	
Input:	*Subset of the input attributes* X_1, X_2,...,X_k
	Subset of the output attribute X_z
	Number of nodes n
Output:	*TRUE/FALSE*

1	$n := 3$
2	$r_c^2 := 0$
3	suffDensity := *TRUE*
4	Create the fuzzy partitions of $[a_1,b_1]$, $[a_2,b_2]$,...,$[a_k,b_k]$ with n nodes
5	*for each* combination $(h_1..h_k)$
6	sum := 0
7	*for* $j = 1 : m$
8	sum := sum + $A_{1h_1}(p_{j1}) \cdot A_{2h_2}(p_{j2}) \cdot ... \cdot A_{kh_K}(p_{jk})$
9	*next* j
10	*if* sum = 0 *then*
11	suffDensity := *FALSE*
12	*exit for*
13	*end if*
14	*next* $(h_1..h_k)$
15	*return* suffDensity

In [2] experiments are performed on a dataset containing oceanographic and surface meteorological data measured from a series of buoys positioned throughout the Equatorial Ocean Pacific (URL: kdd.ics.uci.edu/databases/el_nino/el_nino.data.html). These data are used in the El Nino/Southern Oscillation (ENSO) cycles and consist of the following attributes:

- X_1 = date,
- X_2 = latitude,
- X_3 = longitude,
- X_4 = zonal winds (west < 0, east > 0),
- X_5 = meridional winds (south < 0, north > 0),
- X_6 = relative humidity,
- X_7 = air temperature (Celsius degrees),
- X_8 = sea surface temperature (Celsius degrees).

Table 8.1 Best values of the indices of determinacy for the dependency of the attribute X_8 in the El Nino dataset [2]

X_z	$X_1, ..., X_k$	r_c^2	$r_c'^2$
X_8	X_1, X_7	0.827	0.828
X_8	X_6, X_7	0.341	0.338
X_8	X_4, X_5	0.266	0.266
X_8	X_1, X_6, X_7	0.110	0.111

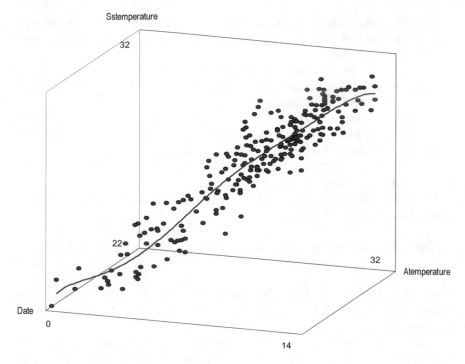

Fig. 8.5 Plot of $X_8 = H(X_1, X_7)$. In red the trend of H^F [2]

In Table 8.1 the best values of the index of determinacy obtained with the multi-dimensional F-transform are shown for dependency between the above attributes.

These results show a dependency of X_8 from X_1 and X_7: in fact, in this case, both the two indices of determinacy are greater than 0.8. The authors, based on results obtained in various tests performed on sample datasets, set 0.8 as threshold value for the two indices of determinacy. In Fig. 8.5 the data points are plotted and the trend of $H^F(X_1, X_7)$ is shown as well.

8.5 Fuzzy Transform Method for Mining Association in the Data

In [1] the multi-dimensional F-transform is applied for mining association in the data.

A mining association in the data in a dataset can be expressed in the form (8.6) and the multi-dimensional discrete inverse F-transform (8.8) can be used to explore this association.

However, unlike to the functions describing dependency between attributes, mining associations are fuzzy functions which establish a correspondence between universes of fuzzy sets.

Let $U_1, \ldots, U_k$ be the domains of k attributes partitioned by fuzzy sets: a mining association functionally joins some fuzzy sets from partitions of $U_1 \ldots U_k$ with fuzzy sets over respective F-transform components.

Let $\{A_{ih_1}, \ldots, A_{ih_i}, \ldots, A_{ih_i}\}$ be an uniform fuzzy partition of the domain U_i of the ith attribute U_i constructed as basic functions of the domain U_i. The fuzzy partition is obtained on the n_i nodes $x_{i1}, \ldots, x_{in_i}$ in the domain U_i.

Each association is supported by two parameters, namely, the degrees of support r and confidence γ defined below.

In [1] the multi-dimensional F-transform is applied in order to discover associations rules in the following form:

$$(X_1 \text{ is } A_{1h_1})\text{AND}(X_2 \text{ is } A_{2h_2})\text{AND} \ldots \text{AND}(X_k \text{ is } A_{kh_k}) \sim^F \text{mean}(X_z) \text{ is } C \tag{8.11}$$

where A_{ih_i}, $i = 1, \ldots, k$, models the meaning of the linguistic expression "approximately x_{h_i}". The corresponding logic clause can be read as "X_i is approximately x_{h_i}".

The label C in the consequent is a linguistic expression given by one of the expressions Sm (small), Me (medium), Bi (big), and eventually combined with one of the following linguistic hedges: Ex (extremely), Si (significantly), Ve (very), empty hedge, ML (more or less), Ro (roughly), QR (quite roughly), VR (very roughly).

The symbol $\sim^F$ expresses an association between the independent and dependent attributes found using the F-transform.

Let O_j, $j = 1, 2, \ldots, m$, be the jth instance of the dataset with component $(p_{j1}, p_{j2}, \ldots, p_{jk}, p_{jz})$. In [1] a membership function of an induced fuzzy set on the set of objects $\{O_1, \ldots, O_m\}$ is defined by considering the antecedent of the hth rule (8.11):

$$A_h(O_j) = A_{1h_1}(p_{j1}) \cdot \ldots \cdot A_{kh_k}(p_{jk}) \tag{8.12}$$

where $A_{ih_i}(p_{ji})$ is the membership degree of the ith attribute of the jth object to the fuzzy set A_{ih_i}.

The following value

$$r = \frac{\text{card}\{O_j \,|\, A_h(O_j) > 0\}}{m} \tag{8.13}$$

is called *degree of support* of the association rule (8.11).

If $F_{h_1 h_2 \ldots h_k}$ is the $(h_1, \ldots, h_k)$th component of the direct F-transform (8.7) and

$$f^F_{n_1 n_2 \ldots n_k}(O_j) = \sum_{h_1=1}^{n_1} \sum_{h_2=1}^{n_2} \cdots \sum_{h_k=1}^{n_k} F_{h_1 h_2 \ldots h_K} \cdot A_{1h_1}(p_{j1}) \cdot \ldots \cdot A_{kh_K}(p_{jk}) \tag{8.14}$$

is the inverse F-transform on the point $(p_{j1}, \ldots, p_{jk})$, in [1] *the degree of confidence* of the association rule (8.11) is defined as

$$\gamma = \sqrt{\frac{\sum_{j=1}^{m} \left(f_{n_1 n_2 \ldots n_k}^{F}(O_j) - F_{h_1 h_2 \ldots h_K} \right) \cdot A_{1h_1}(p_{j1}) \cdot \ldots \cdot A_{kh_K}(p_{jk})}{\sum_{j=1}^{m} \left(p_{jz} - F_{h_1 h_2 \ldots h_K} \right) \cdot A_{1h_1}(p_{j1}) \cdot \ldots \cdot A_{kh_K}(p_{jk})}} \qquad (8.15)$$

The strength of the hth association rule (8.11) is evaluated measuring the degree of support and the degree of confidence (8.14) and (8.15). If both the two parameters are greater or equal to a degree of support threshold and a degree of confidence threshold, respectively, the association is found.

Below we show in the pseudocode, the algorithm that can be used to evaluate if an association (8.11) is found by analyzing a dataset with m objects.

Algorithm: F-transformAssociationRules	
Input:	*Subset of the input attributes $X_1, X_2, \ldots, X_k$*
	Subset of the output attribute X_z
	Threshold values r_{th} and γ_{th}
Output:	r and γ
1	card := 0 // number of object for which $A_h(O_j) > 0$
2	γ_{num} := 0 // numerator of γ in the (8.15)
3	γ_{denom} := 0 // denominator of γ in the (8.15)
4	Calculate the direct F-transform component $F_{h_1 h_2 \ldots h_K}$ by (8.7)
5	*for* j = 1: m
6	Compute $A_h(O_j)$ by (8.12)
7	*if* $A_h(O_j)>0$ *then*
8	card := card+1
9	*end if*
10	Calculate $f_{n_1 n_2 \ldots n_k}^{F}(O_j)$ by (8.14)
11	γ_{num} := $\gamma_{num} + \left(f_{n_1 n_2 \ldots n_k}^{F}(O_j) - F_{h_1 h_2 \ldots h_K} \right) \cdot A_{1h_1}(p_{j1}) \cdot \ldots \cdot A_{kh_K}(p_{jk})$
12	γ_{denom} := $\gamma_{denom} + \left(p_{jz} - F_{h_1 h_2 \ldots h_K} \right) \cdot A_{1h_1}(p_{j1}) \cdot \ldots \cdot A_{kh_K}(p_{jk})$
13	*next* j
14	r := card/m // equation (8.13)
15	$\gamma := \sqrt{\dfrac{\gamma_{num}}{\gamma_{denom}}}$
16	*if* (r $\geq r_{th}$) AND ($\gamma \geq \gamma_{th}$) *then*
17	**return** *'Association found'*, r, γ
18	*else*
19	**return** *'Association not found'*, r, γ
20	*end if*

In [1] this method is applied on about 500 observations in a dataset collected by the Norwegian Public Roads Administration in which air pollution on a specific road is related to traffic intensity and meteorological characteristics. The measures are performed at Alnabru in Oslo (Norway) in the period October 2001–August 2003.

The attributes considered in this dataset are the following:

- $X1$—logarithm of the number of cars per hour
- $X2$—temperature measured 2 m above ground (°C)
- $X3$—wind speed (m/s)
- Xz—hourly values of the logarithm of concentration of NO_2.

The following domains of the four attributes are: $U_1 = [4.127, 8.349]$, $U_2 = [-18.6, 21.1]$, $U_3 = [0.3, 9.9]$, $U_2 = [1.2, 6.0]$. The threshold for the degree of support r is 0.11 and the one for the degree of confidence γ is 0.19. 18 associations are found: the association rule with the best degree of support ($r = 0.34$) and degree of confidence ($\gamma = 0.32$) is the following rule:

$$\left(X_1 \text{ is } A_{1h_1} \right) \text{AND} \left(X_2 \text{ is } A_{2h_2} \right) \text{AND} \left(X_3 \text{ is } A_{3h_3} \right) \sim^F \text{mean}(X_z) \text{is RoBi}$$

where A_{1h_1} is centered on the node $x_{1h_1} = 7$, A_{2h_3} is centered on the node $x_{2h_2} = -1.6$, and A_{3h_3} is centered on the node $x_{3h_3} = 2.1$.

This result can be expressed as: "*if the logarithm of number of cars is approximately 7, the temperature 2 m above ground is approximately −1.6 and the speed of the wind is approximately 2.2 m/s, then the mean concentration of NO_2 is roughly high*".

This method could be also applied for mining association rules in massive datasets partitioning randomly the dataset in subsets and measuring strength of the hth association in each subset. Chapter 12 is dedicated to a discussion on the use of the multi-dimensional fuzzy transform in massive datasets analysis.

References

1. Perfilieva, I., Novàk, V., & Dvoràk, A. (2008). Fuzzy transforms in the analysis of data. *International Journal of Approximate Reasoning, 48*, 36–46.
2. Di Martino, F., Loia, V., & Sessa, S. (2010). Fuzzy transforms method and attribute dependency in data analysis. *Information Sciences, 180*(4), 493–505.
3. Johnson, R. A., & Wichern, D. W. (1998). *Applied Multivariate Statistical Analysis* (773 pp.). London: Prentice-Hall International. ISBN: 0-13-187715-1.

Chapter 9
Fuzzy Transforms in Prevision Analysis

Forecasting techniques are used in various problems and in different application fields: generally, they employ technical statistics to make predictions based on a set of available historical data. Several researchers have proposed soft computing approaches to implement forecasting models, such as fuzzy models, neural networks, and vector quantization techniques. In this chapter, we will discuss the use of the one-dimensional and multi-dimensional discrete F-transforms in data analysis to make predictions.

A significant application of data prediction models concerns time series forecasting: time series is a set of measures of random variables ordered with respect to time. They are used to study the dynamics of a certain phenomenon over time and to predict its future trend.

9.1 Time Series Forecasting

Time series forecasting involves methods for fitting over historical data referring to measures of an observable and using them to predict future observations.

A time series is given by a set of data measured at different times listed in time order.

Let y be the measured output variable, the n measured data are listed in time order as $y(1), y(2),\ldots, y(n)$. Formally, a time series is a function $y: t \in N \to y(t) \in R$ to be known in n time moments. Aim of forecasting is to assess the m future (unknown) values of the output variable $y(n + 1),\ldots, y(n + m)$. In order to estimate the value $y(t + 1)$ of the output variable in the time $t + 1$ are used the last $k + 1$ measured values $y(t), y(t - 1),\ldots, y(t - k)$.

Let $y(t), t = 1, 2,\ldots, T$, be a time series. It can be decomposed by two terms:

$$y(t) = f(t) + r(t) \tag{9.1}$$

© Springer Nature Switzerland AG 2020
F. Di Martino and S. Sessa, *Fuzzy Transforms for Image Processing and Data Analysis*,
https://doi.org/10.1007/978-3-030-44613-0_9

where $f(t)$ is a deterministic part, called a trend, and $r(t)$ is an additional random part called residuals, measuring the error with respect to the trend at the time t. A general model of a stationary time series $y(t)$ can be represented in the form of auto-regressive linear function [1]:

$$y(t) = \alpha_1 y(t-1) + \cdots + \alpha_p y(t-p) + \varepsilon_t \qquad (9.2)$$

where coefficients satisfy some requirements and ε_t is the statistical white noise representing fluctuations in the observations that cannot be explained by the model. The form (9.2) is denoted by Auto-Regressive of order p or AR(p), and it is the simplest representation of the time series $y(t)$, which can be used for its forecast.

Various statistical methods can be used in time series analysis. Generally, statistical time series methods are classified in frequency-domain and time-domain methods. Frequency-domain methods apply frequency analysis approaches as methods based on Fourier analysis, time-domain methods apply auto-regressive models as Bayesian auto-regressive, moving average and ARIMA models in which an auto-regressive linear equation is applied to predict the output variable at the time based on observations from previous time steps.

The Auto-regressive Integrated Moving Average (ARIMA) model generalizes the Auto-Regressive AR(p) models; it is the most used statistical time series prediction model; it can be applied in the data, in order to eliminate the non-stationarity.

ARIMA models are generally denoted ARIMA (p, d, q), where three parameters p, d, q are positive integers in which p is called the order of the auto-regressive model (9.2) or number of time lags, d is called degree of difference and represents the number of non-seasonal differences needed for stationarity, and q is the order of the moving average, representing the number of lagged forecast errors in the prediction equation.

It is proved that in real time series the ARIMA models cannot model sufficiently nonlinear behaviors in the data, so some researchers proposed in literature soft computing and machine learning methods to catch nonlinear behaviors in time series as methods based on multilayer perceptron and convoluted neural networks and vector quantization algorithms, to capture nonlinear components in time series. Recently time series forecasting methods based on the discrete F-transform [2–7] were proposed.

9.2   One-Dimensional Direct F-Transforms in Time Series Analysis

In [3, 4] the unidimensional direct F-transform is applied to assess the trend $f(t)$.

Let $\{y(t), t = 1, 2,..., T\}$ be a time series given by a set of data $y(t)$ measured in regular times. Let $P_T = \{1, 2,..., T\}$, be a set of time moments and let $\{t_1 = 1, t_2,..., t_n = T\}$ be a set of n nodes of the interval $[1, T]$ with $3 \leq n \leq T$. We construct a uniform fuzzy partition of the interval $[1, T]$ with basic functions $A_1,\ldots, A_n$. The dataset given by the time series $\{y(t), t = 1, 2,\ldots, T\}$ is sufficiently dense with respect to this fuzzy partition as $n < T$.

Then we can define the direct unidimensional F-transform of f with components

$$F_k = \frac{\sum_{i=1}^{T} y(t) A_k(t)}{\sum_{i=1}^{T} A_k(t)} \quad k = 1, 2 \ldots, n \tag{9.3}$$

Let $P_k, k = 1,\ldots, n$, be a subset of a subset of P_T given by time moments covered by A_k, i.e.,

$$P_k = \{t \in P_T | A_k(t) > 0\} \tag{9.4}$$

In [3, 4] a model of stationary time series (9.2) is considered in which the time series is decomposed in the sum of its trend and of its residuals being:

$$y(t) = \bigcup_{k=1}^{n} (F_k + r_{tk}) \tag{9.5}$$

where U is given by the max operator and $(r_{1k}, r_{2k},\ldots, r_{Tk})$ is called the kth residual vector of $y(t)$ with respect to A_k, where $t = 1,\ldots, T, k = 1,\ldots, n$.

The residual component r_{tk} is defined as

$$r_{tk} = \begin{cases} y_t - F_k \text{ if } t \in P_k \\ -\infty \quad \text{otherwise} \end{cases} \tag{9.6}$$

The F-transform components $F_1,\ldots, F_n$ is considered to be a new time series with observations F_k and is used to forecast the trend of the original time series $\{y(t), t = 1,\ldots, T\}$.

The trend and the residuals are considered auto-regressive of order p, so the kth component is a linear combination of the p previous components. The trend at the kth time step is assessed by

$$F_k = \alpha_1 F_{k-1} + \alpha_2 F_{k-2} + \cdots + \alpha_p F_{k-p} \quad k = p + 1, \ldots, n \tag{9.7}$$

In [3] $p = 3$ is set. A numerical method and a multilayer perceptron neural network are used to find the optimal values of the coefficients $\alpha_1, \alpha_2, \alpha_3$ [4] also a fuzzy relation model is applied to find the best values of the coefficients. The solutions $\tilde{\alpha}_1, \tilde{\alpha}_2, \tilde{\alpha}_3$ for the three coefficients that minimize the absolute difference between the predicted value and the calculated value for F_n are used to forecast the unknown value F_{n+1} as

$$F_{n+1} = \tilde{\alpha}_1 F_n + \tilde{\alpha}_2 F_{n-1} + \tilde{\alpha}_3 F_{n-3} \tag{9.8}$$

The same process is used to predict the residual components. The value of y at the time $n + 1$ is assessed by (9.5).

Below we schematize in pseudocode the algorithm used in [3, 4]. The user sets the number of nodes n and a time $T_1 < T$. The set of data from the time $T_1 + 1$ to T can be used for testing the performance of the forecasting algorithm.

Algorithm: F-transformTSPrediction Perfilieva et al., 2010a; Perfilieva et al., 2013)	
Input:	*Time series measures $y_1, y_2, \ldots, y_T$*
	Number of nodes n
Output:	*Assessed value of y at the time $T_1+1, \ldots, T$*
1	Create a uniform fuzzy partition of $[1, T_1]$
2	*for* k = 1:n
3	Calculate F_k by (9.3) //calculate the direct F-transform component
4	Calculate the residual vector of F_k $(r_{1k}, r_{2k}, \ldots, r_{Tk})$ by (9.6)
5	*next k*
6	Apply a linear or neural model to find the best value of $\tilde{\alpha}_1, \tilde{\alpha}_2, \tilde{\alpha}_3$
7	Calculate F_{n+1} in (9.8)
8	Repeat this calculus to assess the residuals at the time r_{n+1}
9	*for* t = T_1+1,...T
10	Assess y(t+1) by (9.5)
11	*next t*
12	*return* y(T_1+1),..., y(T)

Two indices are calculated for measuring the performance of the methods given by

$$\text{MAPE} = \frac{1}{T - T_1} \sum_{t=T_1}^{T} \left| \frac{y(t) - \tilde{y}(t)}{y(t)} \right| \tag{9.9}$$

and

$$\text{SMAPE} = \frac{1}{T - T_1} \sum_{t=T_1}^{T} \left| \frac{y(t) - \tilde{y}(t)}{y(t) + \tilde{y}(t)} \right| \tag{9.10}$$

Comparison tests are performed in [4] with other time series fuzzy models and with the ARIMA model. In Table 8.1 the results obtained on the well-known Alabama time series relating to the enrollment of students from the University of Alabama [8] are shown (Table 9.1).

The results in Table 9.1 show that the F-transform time series algorithm in [4] has better performances with respect to those obtained by using the other fuzzy models and the ARIMA model.

Table 9.1 MAPE index measured comparing time series in "Alabama" [4]

Model	MAPE(%)
Song and Chissom [8]	3.18
ARIMA (0,1,1) (0,1,2)	2.1
Hwang et al. [13] and Sah and Degtiarev [14]	5.8
Huarng [15]	1.5
	1.0
Perfilieva et al. [4]	0.4

9.3 Fuzzy Forecasting Analysis: The Wang and Mendel Method

Let $x_1, x_2, \ldots, x_k$, be a set of input numerical variables and let y be the output variable with $x_1, x_2, \ldots, x_k, y \in R$. In forecasting analysis at least one of the input variables is the period in which the measurement was made.

Let $\left(x_1^{(j)}, x_2^{(j)}, \ldots, x_k^{(j)}, y^{(j)} \right)$ be the jth measure of the input and output variables with $j = 1, \ldots, N$. Our aim is to assess the future value of the output variable y by extracting a mapping of the function $y = f(x_1, x_2, \ldots, x_k)$ defined in R^k from the numerical dataset formed by the N measures $\left(X_1^{(j)}, X_2^{(j)}, \ldots, X_k^{(j)}, y^{(j)} \right) j = 1, 2, \ldots, N$.

Time series is a specification of this dataset. For example, we can model a stationary time series (9.2) $y(t) = \alpha_1(t - 1) + \alpha_2(t - 2) + \cdots + \alpha_2(t - k) + \varepsilon_t$ by setting: $x_1 = y(t - 1), x_2 = y(t - 2), \ldots, x_k = y(t - k)$.

The well-known machine learning Wang and Mendel algorithm [9] generates a fuzzy rule set from this numerical dataset: this method can approximate any real continuous function defined on a compact set.

Let $[x^{i-}, x^{i+}]$, $i = 1, 2, \ldots, n$, be the domain of the ith input variable and $[y^-, y^+]$ the domain of the output variable.

In Wang and Mendel method, a fuzzy partition of the domain $[x^{i-}, x^{i+}]$ of any input variable, with $i = 1, \ldots, k$, is created. The domain $[x^{i-}, x^{i+}]$ is partitioned in $2M_i + 1$ *regions*, where the value M_i is made by the expert and can be different for each variable. To any region is assigned s fuzzy sets which are labeled $SM_i, \ldots, S1, CE, BM_i, \ldots, B1$ (Small $M_i, \ldots,$ Small 1, Center, Big 1, Big M_i), respectively. Similarly, the domain $[y^-, y^+]$ of the output variable y is partitioned into $2M_y + 1$ fuzzy sets.

In the next step the N data vectors $(x_1^{(j)}, x_2^{(j)}, \ldots, x_k^{(j)}, y^{(j)})$, $j = 1, \ldots, N$, are analyzed to construct N fuzzy rules. The jth fuzzy rule is constructed by assigning the ith input variable to the fuzzy set of the ith fuzzy partition.

In Fig. 9.1 an example of construction of two fuzzy rules from two data vectors is shown. In this example, a fuzzy partition of $[x^{1-}, x^{1+}]$ composed of 7 fuzzy sets ($M_1 = 3$) is created and a fuzzy partition of $[x^{2-}, x^{2+}]$ composed of 13 fuzzy sets ($M_2 = 6$) is created. The domain of the output variable $[y^-, y^+]$ is partitioned in 7 fuzzy sets ($M_y = 3$).

The two fuzzy rules constructed are the following:

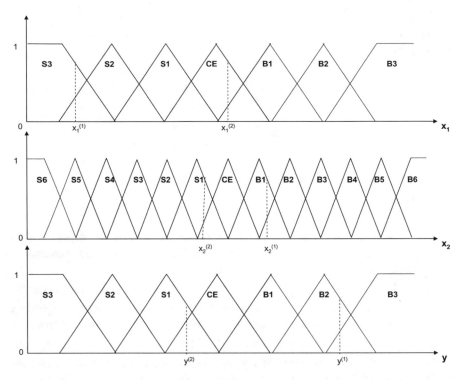

Fig. 9.1 Example of construction of fuzzy rules by Wang and Mendel method

$r^{(1)}$: IF (x_1 is S3) AND (x_2 is B1) THEN y is B2
$r^{(2)}$: IF (x_1 is CE) AND (x_2 is S1) THEN y is S1.

The combination of the fuzzy partitions of the domains of the input variable forms a *rule box framework* whose cells (the rule boxes) represent a combination of the fuzzy sets of the input variables. Of course, the finer the fuzzy partitions, the greater the resolution of the rule box. In Fig. 2 the rule box framework obtained using the fuzzy partitions of the two input variables in the example is shown (Fig. 9.2).

After constructing the N fuzzy rules, each cell is empty or filled with a fuzzy rule, all the duplication rules (antecedent and consequent identical) are merged and all the conflicting rules are analyzed. Two rules are in conflict if they are the same antecedent, that is, they are inserted in the same cell as the rule box, and different consequence. To solve this conflict is considered only the rule with greater strength, calculated as

$$D\big(r^{(j)}\big) = m\big(x_1^{(j)}\big) \cdot m\big(x_2^{(j)}\big) \cdot \ldots \cdot m\big(y^{(j)}\big) \cdot m^{(j)} \tag{9.11}$$

Fig. 9.2 Example of rule box framework used in the Wang and Mendel model

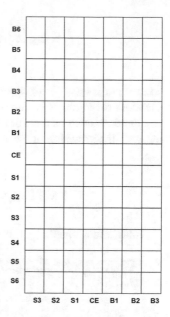

where $m\left(x_i^{(j)}\right)$, $i = 1,\ldots,$ k, and $m(y^{(j)})$ are the membership degree of $x_i^{(j)}$ and $y^{(j)}$ to its related fuzzy sets and $m^{(j)}$ is the *belief of usefulness* of the jth rule assigned by the expert, respectively.

Subsequently, filled rule boxes of a same row or column with identical consequent, they are merged in a "or" fuzzy rule using the or connective.

After constructing the final fuzzy rule set composed of R fuzzy rules, for each rule the membership degree to the fuzzy set in the consequent is calculated. If r_h is the hth fuzzy rule, O^h is the fuzzy set in its consequent and I_i^h is the fuzzy set of the fuzzy partition of the domain of the ith input variable, with $i = 1,\ldots,$ k, the membership degree to O^h is computed by

$$m_{O^h}^h = m_{I_1^h}(x_1) \cdot m_{I_2^h}(x_2) \cdot \ldots\ldots \cdot m_{I_k^h}(x_k) \tag{9.12}$$

The assessed value of the variable y is given by the defuzzification formula:

$$\tilde{y} = \frac{\sum_{h=1}^{R} m_{O^h}^h \bar{y}^h}{\sum_{h=1}^{R} m_{O^h}^h} \tag{9.13}$$

where R is the number of rules and $\bar{y}^h$ is the center value of the region of the fuzzy set O^k with the smallest absolute value between all the points with a membership degree to O^k equal to 1.

Substituting the degree $m_{O^k}^k$ in (9.13) with its value in (9.12) we obtain:

$$\tilde{f}(x_1, x_2, \ldots, x_k) = \tilde{y} = \frac{\sum_{h=1}^{R}\left[\bar{y}^k \prod_{i=1}^{k} m_{l_i}^h(x_i)\right]}{\sum_{h=1}^{R}\left[\prod_{i=1}^{K} m_{l_i}^h(x_i)\right]} \tag{9.14}$$

where $\tilde{f}(x_1, x_2, \ldots, x_k)$ approximates the function $f\ (x_1, x_2,\ldots, x_k)$. Wang and Mendel proved that it is a universal approximator from a compact set of R^k.

An error index as the Root Mean Square Error index (RMSE) can be applied to measure the precision of the approximation. It is given by

$$\text{RMSE} = \sqrt{\frac{\sum_{j=1}^{N}\left(\tilde{f}(x_1^{(j)}, \ldots, x_k^{(j)}) - y^{(j)}\right)^2}{N}} \tag{9.15}$$

Wang and Mendel algorithm is schematized in the pseudocode below.

Algorithm: WangMendel (Wang & Mendel, 1992)	
Input:	*Dataset $(x_1^{(j)}, x_2^{(j)}, \ldots, x_k^{(j)}, y^{(j)})\ j = 1,2,\ldots,N$*
Output:	*Final R fuzzy rules*
	RMSE
1	*for* i = 1:k
2	Partition the domain $[x^{i-}, x^{i+}]$ in M_i regions
3	For any region create the corresponding fuzzy set
4	*next* i
5	Partition the domain $[y^-, y^+]$ in M_y regions
6	For any region of $[y^-, y^+]$ create the corresponding fuzzy set
7	*for* j = 1:N
8	Construct the j*th* fuzzy rule
9	*next j*
10	Delete the duplication rules
11	Solving conflicting rules by using (9.11)
12	Create the "or" fuzzy rules obtaining R fuzzy rules
13	Store the R fuzzy rules
14	*for* j = 1:N
15	Calculate $\tilde{f}(x_1, x_2, \ldots, x_k)$ by (9.14)
16	RMSE:=RMSE+$(\tilde{f}(x_1^{(j)}, x_2^{(j)}, \ , x_k^{(\ )}) - y^{(j)})^2$
17	*next j*
18	RMSE := $\sqrt{\dfrac{RMSE}{N}}$
19	*return* RMSE

In [9] this algorithm has been applied for forecasting on the well-known Chaotic Mackey Glass time series. The authors have shown that the results obtained using this method are comparable with that one obtained by using a multilayer perceptron neural network model.

9.4 Multi-dimensional F-Transform for Forecasting in Data Analysis

As the function approximator (9.14) constructed by using the Wang and Mendel method, the inverse multi-dimensional F-transform can be used as function approximator applied in forecasting analysis.

Let $(x_1^{(j)}, x_2^{(j)}, \ldots, x_k^{(j)}, y^{(j)})$ be the jth measure of the input and output variables with $j = 1, \ldots, N$.

Let $[x^{i-}, x^{i+}]$, $i = 1, 2, \ldots, n$, be the domain of the ith variable. Let $\{A_{i1}, A_{i1}, \ldots, A_{i1k}\}$ be an uniform fuzzy partition of $[x^{i-}, x^{i+}]$ composed by n_i fuzzy sets. If the set of data is sufficiently dense with respect to the fuzzy partitions, it is possible to construct the discrete multiple direct Fuzzy Transform with components:

$$F_{h_1 h_2 \ldots h_k} = \frac{\sum_{j=1}^{N} y^{(j)} \cdot A_{1_{h_1}}(x_1^{(j)}) \cdot \ldots \cdot A_{k_{h_n}}(x_k^{(j)})}{\sum_{j=1}^{N} A_{1_{h_1}}(x_1^{(j)}) \cdot \ldots \cdot A_{k_{h_n}}(x_k^{(j)})} \qquad (9.16)$$

The inverse F-transform is given by

$$f_{n_1 n_2 \ldots n_K}^F (x_1^{(j)}, x_2^{(j)}, \ldots, x_k^{(j)}) = \sum_{h_1=1}^{n_1} \sum_{h_2=1}^{n_2} \cdots \sum_{h_K=1}^{n_k} F_{h_1 h_2 \ldots h_K} \cdot A_{1_{h_1}}(x_1^{(j)}) \cdot \ldots \cdot A_{k_{h_K}}(x_1^{(j)}) \qquad (9.17)$$

The (9.16) is used in [2] to approximate the map of the function $y = f(x_1, x_2, \ldots, x_k)$.

Compared to the time series forecasting method proposed in [3], the multi-dimensional F-transform forecasting method proposed in [2] can be applied only by appropriately choosing the set of fuzzy partition of k domains in order to guarantee sufficient data density. On the other hand, this method generalizes the use of the F-transform in forecasting analysis, taking into account dependency of the observable y not only from the time in which it was observed but also from additional input variables that participate in the conditions in which the measure has been completed.

It is necessary to point out that a constraint similar to that of the sufficient density of data points with respect to fuzzy partitions is also present in the algorithm of Wang and Mendel in which to obtain acceptable performance it is necessary that the rule boxes are filled. The constraint of sufficient data density appears in different ways in any machine learning algorithm used as a function approximator.

In [2] an approach is proposed to control the sufficient density of the data with respect to the fuzzy partitions in which two indices are measured in order to evaluate the forecast accuracy. The error of the approximated map is calculated by using the Root Mean Square Error (RMSE) and the Mean Absolute Deviation Mean (MADMEAN), given by

$$\text{RMSE} = \sqrt{\frac{\sum_{j=1}^{N} \left(f_{n_1 n_2 \ldots n_k}^F (x_1^{(j)}, \ldots, x_k^{(j)}) - y^{(j)} \right)^2}{N}} \qquad (9.18)$$

and

$$\text{MADMEAN} = \frac{\sum_{j=1}^{N} \left| f_{n_1 n_2 \ldots n_k}^F (x_1^{(j)}, \ldots, x_k^{(j)}) - y^{(j)} \right|}{\sum_{j=1}^{N} y^{(j)}} \qquad (9.19)$$

The choice of the MADMEAN index is justified by the results obtained in [10] in which the authors show the advantage of the use of the MADMEAN index respect to the well-known MAPE and MAD indices to evaluate the accuracy of a forecasting model.

The algorithm used in [2] is an iterative algorithm in which initially coarse-grained fuzzy partitions are created ($n = 3$); the direct (9.11) and inverse (9.12) multiple fuzzy transforms are used and the RMSE and MADMEAN indexes are calculated. If RMSE is greater or equal a RMSE threshold and MADMEAN is greater or equal to a MADMEAN threshold, the process stops and the direct F-transform components are stored, else, more fine-grained fuzzy partitions are created. In each cycle, the algorithm controls that the data are sufficiently dense with respect to the fuzzy partitions. If they are not, the algorithm ends and the forecast has not been made.

Below the algorithm used in [2] is shown. The algorithm returns the value of the variable *FoundPrediction* whose value is *True* if the prediction is performed, *False* otherwise. If the prediction is performed the direct F-transform components are stored in order to give the value of the output variable by (9.17).

Algorithm: MultipleF-transformPrediction (Di Martino & Sessa, 2011)	
Input:	Dataset $x_1^{(j)},...,x_k^{(j)}, y^{(j)})_j$ $= 1,2,...,N$ RMSE threshold RTh MADMEAN threshold MTh
Output:	*Direct Multiple F-transform*

20	n:=3
21	FoundPrediction := False
22	StopCycle:=False
23	*for* i = 1:k
24	Create a uniform partition of n fuzzy sets of $[x^{i^-}, x^{i^+}]$,
25	*next* i
26	*while* StopCycle = False
27	*if* the data are suff. Dense with respect to the fuzzy partitions *then*
28	*for each* combination $(h_1, h_2,..., h_k)$
29	Calculate $F_{n_1 \cdots n_k}$ by (9.16)
30	*next*
31	RMSE := 0
32	MADMEAN := 0
33	SUMY := 0
34	*for j = 1,...,N*
35	Calculate $f_{n_1 n_2 ... n_k}^F (x_1^{(j)}, x_2^{(j)},..., x_k^{(j)})$ by (9.17)
36	RMSE := RMSE + $(f_{n_1 n_2 ... n_k}^F (x_1^{(j)}, x_2^{(j)},..., x_k^{(j)}) - y^{(j)})^2$
37	MADMEAN := MADMEAN + $\left| f_{n_1 n_2 ... n_k}^F (x_1^{(j)}, x_2^{(j)},..., x_k^{(j)} - y^{(j)}) \right|$
38	SUMY := SUMY + $y^{(j)}$
39	*next j*
40	RMSE := $\dfrac{RMSE}{N}$
41	MADMEAN := MADMEAN/SUMY
42	*if* (RMSE > RTh) *OR* (MADMEAN > MTh) *then*
43	n:= n+1
44	*else*
45	Store the multiple direct F-transform components
46	StopCycle:=TRUE
47	FoundPrediction := True
48	*end if*
49	*else*
50	StopCycle:=TRUE
51	*end if*
52	*end while*
53	*return* FoundPrediction

In [2] comparison tests of the multi-dimensional F-transform with the Wang and Mendel and the Local Linear Wavelet Neural Network algorithms are performed on the Mackey glass time series generated by the following differential equation:

$$\frac{dy}{dx} = \frac{0.2 \cdot y(x - \tau)}{1 + y^{10}(x - \tau)} - 0.1 \cdot y(x) \qquad (9.20)$$

This differential equation has a chaotic behavior if $\tau \geq 17$. In these tests, the function $y(x)$ is approximated by considering the sequence $y(x-6)$, $y(x-5)$,..., $y(x-1)$. Each input/output data point is given by the ennuple $(x_1, x_2, x_3, x_4, x_5, x_6, y)$. A training set of 700 data points to predict the next 300 points are used performing a partition of 35 fuzzy sets for each variable. Various fuzzy partitions are used by applying the multi-dimensional F-transform and the Wang and Mendel methods with overall 108, 210, 246, 300 fuzzy sets are used corresponding to partition of 18, 35, 41, 50 fuzzy sets of each input variable, respectively. A local Wavelet Neural Network with 3 layers, 6 input nodes, 10 hidden nodes, and 1 output node is used.

The RMSE threshold and the MADMEAN threshold are set to 0.3 and 0.5, respectively, applying the multi-dimensional F-transform algorithm.

Table 9.2 shows the RMSE and MADMEAN obtained in any experiment setting $\tau = 30$.

Figures. 9.3, 9.4, and 9.5 show the plot of the predicted value of $y(x)$ for x between 701 and 1000 obtained by using the three methods, where the multi-dimensional F-transform and the Wang and Mendel algorithm are performed considering 300 fuzzy sets.

Other experiments are performed in [2] applying the multi-dimensional F-transform algorithm to forecast the value of the North Atlantic Oscillation (NAO) index.

The NAO index [11, 12] is a climatic index given by the difference of sea-level pressure between two stations situated close to the centers of Icelandic Low and Azores High. A positive NAO index (NAO$_+$) reflects a warm and wet winter in northern Canada and Greenland, in the eastern US and in the west Europe. A negative

Table 9.2 MADMEAN and RMSE indices measured comparing the Multi-dimensional F-Transform, Wang and Mendel and LLWNN methods for the time series "Mackey glass" with $\tau = 30$ [2]

Prediction method	RMSE	MADMEAN
F-Transform with a108 fuzzy sets	0.08285	1.17952
W & M with 104 fuzzy sets	0.08254	1.18453
F-Transform with 210 fuzzy sets	0.03193	0.64712
W & M with 210 fuzzy sets	0.03251	0.64568
F-Transform with 246 fuzzy sets	0.02312	0.30201
W & M with 246 fuzzy sets	0.02623	0.34577
F-Transform with 300 fuzzy sets	0.01743	0.30205
W & M with 300 fuzzy sets	0.02042	0.28372
LLW with 6 input, 10 hidden and 1 output neurons	0.02474	0.56476

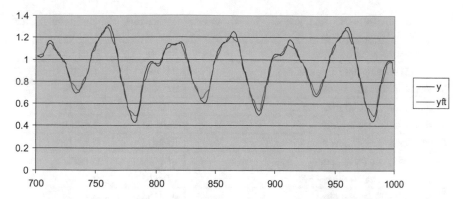

Fig. 9.3 Prediction of the chaotic Mackey glass time series with $\tau = 30$ using a training set of 700 points and the multi-dimensional F-Transform method with 300 fuzzy sets [2]

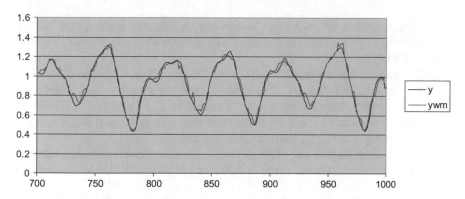

Fig. 9.4 Prediction of the chaotic Mackey glass time series with $\tau = 30$ using a training set of 700 points and the Wang and Mendel algorithm with 300 fuzzy sets [2]

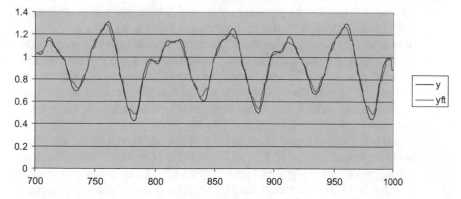

Fig. 9.5 Prediction of the chaotic Mackey glass time series with $\tau = 30$ using a training set of 700 points and a Local Linear Wavelet Neural Network [2]

Table 9.3 MADMEAN   and RMSE   indices measured comparing the multi-dimensional F-Transform, Wang and Mendel and LLWNN methods for the monthly NAO index time [2]

Prediction method	RMSE	MADMEAN
F-Transform with 108 fuzzy sets	0.07409	1.71221
W & M with 104 fuzzy sets	0.07499	1.76004
F-Transform with 210 fuzzy sets	0.02825	0.64717
W & M with 210 fuzzy sets	0.02861	0.64286
F-Transform with 46 fuzzy sets	0.01975	0.45817
W & M with 246 fuzzy sets	0.02050	0.45921
F-Transform with 300 fuzzy sets	0.01492	0.34295
W & M with 300 fuzzy sets	0.01584	0.35954
LLW with 6 input, 10 hidden, and 1 output neurons	0.02448	0.59472

NAO index reflects a cold winter in these areas and storms track southerly toward the Mediterranean Sea. This brings increased storm activity and rainfall to southern Europe and North Africa.

This experiment uses a dataset of 712 mean monthly NAO index values derived by daily measures from January 1950 to April 2009. The monthly NAO index at the time t is assessed by considering this value at the times $y(t-6)$, $y(t-5)$,…, $y(t-1)$. The dataset is partitioned in a training set of 500 points used to predict the next 212 points; 300 fuzzy sets (50 for each variable) are used.

In Table 9.3 the RMSE and MAD MEAN error values using different values of the number of fuzzy sets used in the F-Transform and Yang and Mendel methods are shown.

Also in this experiment the three algorithms provide comparable performances. The multi-dimensional F-transform algorithm provides the best results in tests where a total of 246 and 300 fuzzy sets have been used.

In Figs. 9.6, 9.7, and 9.8 the plot of the predicted NAO index values obtained by using the three methods with respect to the measured values are shown.

The results of these experiments show that multi-dimensional F-transform provides performance comparable to that one provided by the Wang and Mendel algorithm and by a Local Linear Wavelet Neural Network. Moreover, the former method has the advantage of verifying the density of the data with respect to the fuzzy partitions, determining which fuzzy partitions generate results that are performance acceptable.

The weak point of the multi-dimensional F-transform method is in the choice of the threshold values for the RMSE and MADMEAN indices. These thresholds can be determined by checking for which dimensionality of the fuzzy partitions the index value reaches a plateau where finer partitioning does not produce significant improvements in performance.

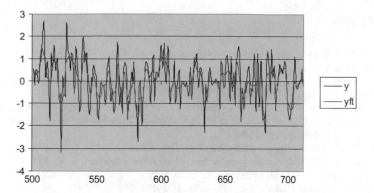

Fig. 9.6 Prediction of the monthly NAO index time series using a training set of 500 points and the multi-dimensional F-Transform method with 300 fuzzy sets [2]

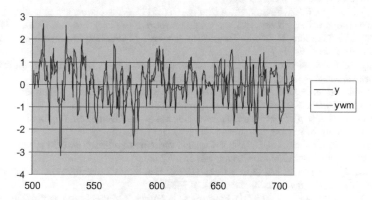

Fig. 9.7 Prediction of the monthly NAO index time series using a training set of 500 points and the Wang and Mendel algorithm with 300 fuzzy sets [2]

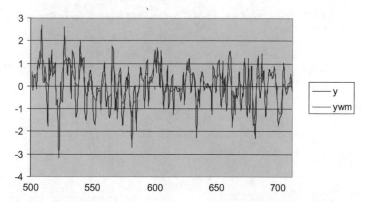

Fig. 9.8 Prediction of the monthly NAO index time series using a training set of 500 points and a Local Linear Wavelet Neural Network [2]

References

1. Wold, H. (1938). *A Study in the Analysis of Stationary Time Series* (p. 236). Stockholm: Almqvist and Wiksel ets.
2. Di Martino, F., Loia, V., & Sessa, S. (2011). Fuzzy transforms method in prediction data analysis. *Fuzzy Sets and Systems, 180,* 146–163.
3. Perfilieva, I., Yarushkina N., & Afanasieva T. (2010). Time series analysis by discrete F-transform. In *Conference Paper in IEEE International Conference on Fuzzy Systems.* https://doi.org/10.1109/fuzzy.2010.5584264.
4. Perfilieva, I., Yarushkina, N., Afanasieva, T., & Romanov, A. (2013). Time series analysis using soft computing methods. *International Journal of General Systems, 42*(6), 687–705.
5. Novak, V., Perfilieva, I., Holcapek, M., & Kreinovich, V. (2014). Filtering out high frequencies in time series using F-transform. *Information Sciences, 274,* 192–209.
6. Novak, V., & Perfilieva, I. (2015). Forecasting direction of trend of a group of analogous time series Using F-transform and fuzzy natural logic. *International Journal of Computational Intelligence Systems, 8*(15), 15–28.
7. Lee, W. J., Jung, H.-Y., Yoon, J. H., & Choi, S. H. (2017). A novel forecasting method based on F-transform and fuzzy time series. *International Journal of Fuzzy Systems, 19*(6), 1793–1802.
8. Song, Q., & Chissom, B. (1993). Forecasting enrollments with fuzzy time series. Part I. *Fuzzy Sets and Systems, 54,* 1–9.
9. Wang, L. X., & Mendel, J. X. (1992). Generating fuzzy rules by learning from examples. *Transactions on Systems, Man, and Cybernetics., 22,* 1414–1427.
10. Kolassam, S., & Schutz, W. (2007). Advantages of the MAD/MEAN ratio over the MAPE. *Foresight: The International Journal of Applied Forecasting, 6,* 40–43.
11. Hurrell, J. W. (1995). Decadal trends in the North Atlantic Oscillation: Regional temperatures and precipitation. *Science, 269,* 676–679.
12. Hurrell, J. W., & van Loon, H. (1997). Decadal variations in climate associated with the North Atlantic Oscillation. *Climatic Change, 36,* 301–326.
13. Hwang, J. R., Chen, S. M., & Lee, C. H. (1998). Handling forecasting problems using fuzzy time series. *Fuzzy Sets and Systems, 100,* 217–228.
14. Sah, M., & Degtiarev K. Y. (2004). Forecasting Enrollment Model Based on First-Order Fuzzy Time Series. In *Proceedings of the International Conference on Computational Intelligence, Istanbul,* 17–19 December 2004 (Vol. 1, pp. 375–378). World Academy of Science, Engineering and Control.
15. Huarng, K. (2006). The application of neural networks to forecast fuzzy time series. *Physica A, 336,* 481–491.

Chapter 10
Fuzzy Transforms Applied in Seasonal Time Series Analysis

Seasonal time series are time series in which is detected the presence of *seasonality*, a phenomenon given by generally repetitive fluctuations in the time occurred at specific regular time intervals less than a year, such as weekly, monthly, or quarterly. Seasonality may be caused by various factors, such as weather, vacation, and holidays and consists of periodic, repetitive, and generally regular and predictable patterns in the levels of a time series.

The analysis of seasonal time series provides a better understanding of the behavior of seasonal variations in time series and to forecast future seasonal fluctuations.

Seasonal time series are present in many fields such as climatology, meteorology, finance, and marketing. Seasonal time series forecasting attracts the interest of many actors and organizations who try to predict future regular fluctuations over time of a specific phenomenon related, for example, to climate change or market trends. Thus seasonal time series forecasting applications can be valuable support tools for organizations to plan future actions.

F-transform techniques can be applied as a regression function in order to predict the trend and the seasonal fluctuations of time series. This regression function is given by the discrete inverse F-transform: the F^1-transform represents a refinement of the F-transform for approximating this regression function and can be used to improve the forecasting performances.

10.1 Seasonal Time Series

Time series can be classified based on their fluctuations over time. Seasonal time series have fluctuations generally with similar intensity in the same periods. For example, the warmest daily temperature is recorded periodically in the summer season.

In cyclical time series oscillations are present in not fixed periods. For example, the wettest day in a geographical area can often be recorded in the autumn season, but sometimes it occurs also in other seasons of the year.

© Springer Nature Switzerland AG 2020
F. Di Martino and S. Sessa, *Fuzzy Transforms for Image Processing and Data Analysis*,
https://doi.org/10.1007/978-3-030-44613-0_10

In irregular time series short-term irregular oscillations are present. Normally they are caused by a stationary stochastic process.

A cyclic pattern exists when data exhibit rises and falls that are not of fixed period. The duration of these fluctuations is usually of at least 2 years: think of business cycles which usually last several years, but where the length of the current cycle is unknown beforehand.

If these fluctuations are repeated in a similar manner in the same period, then the pattern is seasonal. In general, the average length of cycles is longer than the length of a seasonal pattern and the magnitude of cycles tends to be more variable than the magnitude of seasonal patterns.

A seasonal time series can be decomposed into three patterns:

Trend $\rightarrow$ a general systematic linear or (most often) nonlinear component that changes over time and does not repeat,
Seasonality $\rightarrow$ a general systematic linear or (most often) nonlinear component that changes over time and does repeat,
Noise $\rightarrow$ a component representing random fluctuations in the time series.

The seasonal period, (say) s, defines the number of observations of which a seasonal cycle is composed: for example, $s = 1$ for the seasonal period is a year, $s = 12$ for monthly series, $s = 4$ for quarterly series.

The simplest models used to approximate a seasonal component are the deterministic models in which the seasonality is given by the same function. Unfortunately, seasonal time series generally do not behave in a deterministic way.

A more complex model to approximate seasonal time series is to assume that seasonality is not the same, but changes and its evolution is stationary, that is mean and variance are constants.

Finally, we can model seasonality by using a non-stationary process, in which seasonality evolves in a not stationary mode.

Various statistical methods can be used to analyze seasonal time series. The most used statistical prediction model is the Seasonal Auto-regressive Integrated Moving Average (SARIMA) model [1]: it is an extension of the ARIMA model that predicts seasonal components in univariate data. A SARIMA model is decomposed into two parts: a non-seasonal part used to model the trend component and a seasonal part used to model the seasonal component.

Likewise ARIMA, SARIMA cannot model accurately nonlinear behaviors in seasonal time series; in contrast, machine learning modeling approaches based on neural network architectures as Long Short-Term Neural Networks or based on vector quantization approaches can be used to detect nondeterministic and nonlinear changes in seasonal components. However, these models are computationally complex and in many cases, it is preferable to use simpler computation models, which provide acceptable performances.

In the next section, a method based on the discrete F-transform proposed in Novak et al. [2] is shown. It aims to capture trends in seasonal time series: in Sects. 10.3 and 10.4 two algorithms are discussed and proposed in Di Martino and Sessa [3] and Di Martino and Sessa [4], respectively. In the first algorithm, the unidimensional discrete

F-transform is used to model the seasonal component, in the latter the unidimensional discrete F^1-transform is used to capture the seasonal component behavior.

10.2 F-Transform Technique to Remove Seasonal Components and Noise from Time Series

F-transform can be applied to smooth a time series in order to assess its trend; the smoothing phase is necessary to eliminate the fluctuations and the noise present in the data. Novak et al. [2] propose a technique based on F-transform to detect the trend in seasonal time series. Starting from the definition of Trend Cycle (TC) as the component of a time series with variations of low frequency, the authors apply F-transform to extract the trend in a seasonal time series as a smooth function, filtering the seasonal component S given by a mixture of periodic function and reducing the noise component R.

Formally, considering a seasonal time series as a stochastic process, the time series can be considered as the sum of three components as

$$y_0(t, \omega) = TC(t) + S(t) + R(t, \omega) \quad \omega \in \Omega \tag{10.1}$$

where Ω is a probabilistic space.

The seasonal component $S(t)$ is given by a sum of r complex periodic functions (the sub-components) defined in the complex field:

$$S(t) = \sum_{i=1}^{r} P_j e^{i(\lambda_j t + \varphi_j)} \tag{10.2}$$

where P_j is the amplitude, λ_j is the angular frequency, and φ_j the phase shift of the jth sub-component.

The random noise $R(t, \omega)$ is assumed to be a stationary stochastic process with zero mean and finite variance. In Novak et al. [2] the noise component is approximated with the following function:

$$R(t, \omega) = \xi(\omega) e^{i(\lambda t + \varphi)} \tag{10.3}$$

where $\xi(\omega)$ is a random function with zero mean value and λ is a real number.

In Novak et al. [2] an unidimensional F-transform is applied in order to remove the seasonal high frequency sub-components. An uniform fuzzy partition of $[0, T]$ formed by the basic functions $A_0, \dots, A_n$ is created, where $A_k, k = 0, \dots, n$, is given by a triangular fuzzy set with width $2h$ and h is the distance between two consecutive nodes.

The nodes are given by $c_0 = 0, c_1 = dT, \dots, c_{k-1} = (k-1)dT, c_k = kdT, \dots, c_n = ndT = T$ with $d = 1/n$. We have that $h = dT$.

The trend component is modeled by using the unidimensional F-transform as

$$\widetilde{T}_C(t) = \sum_{h=1}^{n} F_h \cdot A_h(t) \tag{10.4}$$

The authors show that the inverse F-transform (10.4), constructed by using these triangular basic functions, can approximate the trend component TC, reducing the seasonal component and the noise.

In Novak et al. [2] comparison tests are performed applying the F-transform to detect TC component in real time series: the authors compare this method with the Season-Trend-Loess (STL) regression method [5], the Butterworth filter [6, 7], and the Singular Spectrum Analysis (SSA) method [8, 9].

The following three real time series are used:

- monthly Canadian unemployment figures in the years 1956–1975,
- monthly number of slaughtered pigs in Victoria in the years 1980–1995,
- monthly accidental deaths in the USA in the years 1973–1978.

The following distance between consecutive nodes are used by applying the F-transform algorithm: $h = 34$ for Canadian unemployment figures, $h = 31$ for monthly slaughtered pigs, $h = 12$ for monthly accidental deaths.

In Figs. 10.1, 10.2, and 10.3 the three time series and the TC component obtained by using the four methods are plotted, respectively.

These results show that the best results are obtained using the F-transform, STL and SSA methods. These three methods provide similar trends, however, the F-transform method is much easier to use than STL and SSA algorithms and has the best computational complexity.

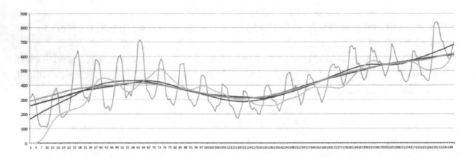

Fig. 10.1 Trends of the Canadian unemployment figures time series by using F-transform (green line), STL method (orange line), Butterworth filter (light blue), and SSA-method (yellow line), in Novak et al. [2]

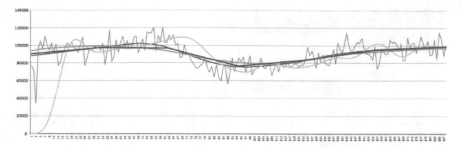

Fig. 10.2 Trends of the monthly number of slaughtered pigs time series by using F-transform (green line), STL method (orange line), Butterworth filter (light blue), and SSA-method (yellow line), in Novak et al. [2]

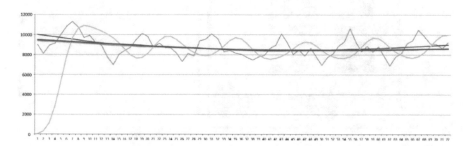

Fig. 10.3 Trends of the monthly accidental deaths time series by using F-transform (green line), STL method (orange line), Butterworth filter (light blue), and SSA-method (yellow line), in Novak et al. [2]

10.3   Seasonal Time Series Fuzzy Transform Forecasting

In Di Martino and Sessa [3] a seasonal time series forecasting method based on the unidimensional F-transform is proposed and the authors call this method Time Series Seasonal F-transform (TSSF).

This algorithm evaluates the trend of the time series, then the approximate trend is subtracted from the time series (detrending operation). The detreated time series is partitioned in seasonal subsets and finally, the unidimensional F-transform is applied to any subset and the direct F-transform components are stored.

A heuristic iterative process is implemented to assess the direct F-transform components from any seasonal subset. Initially, a coarse-grained fuzzy partition is created ($n = 3$) and the algorithm verifies if the subset of data is sufficiently dense with respect to the partition. If this is true, then the direct F-transform components and the inverse F-transform at any point are calculated in order to evaluate the MADMean index. If this index is less or equal to a prefixed threshold, then the process stops and the direct F-transform components are stored, else a fine-grained fuzzy partition is created ($n := n + 1$) and the process is iterated.

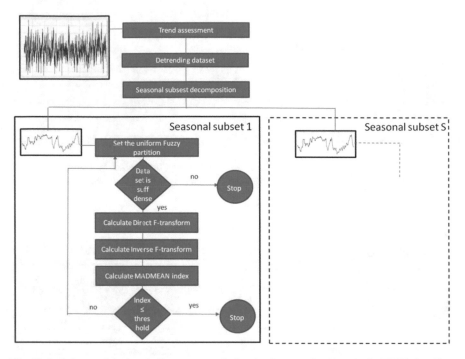

Fig. 10.4 Schema of the heuristic process applied to the S seasonal subsets in the TSSF algorithm

The process also ends if the data are not sufficiently dense with respect to the fuzzy partition: in this case it is not possible to predict the future behavior of the seasonal time series.

Figure 10.4 schematizes the heuristic process applied to any seasonal subset in the TSSF method.

This algorithm is performed for all the S seasonal subsets: of course, the size $n(1)$, ..., $n(S)$ of each fuzzy partition can be different since the density of points can vary from seasonal subset to seasonal subset.

The seasonal data subset is composed by M_s pairs, expressing the fluctuation measures of the parameter y_0 at different times: $\left(t^{(1)}, y^{(1)}\right), \left(t^{(2)}, y^{(2)}\right) \ldots \left(t^{(M_s)}, y^{(M_s)}\right)$, where $y^{(i)}$ is given by the original measure $y_0^{(j)}$ at the time $t^{(i)}$ minus the trend calculated at this time. To assess the time series behavior in a time t, where t belongs to the season s, the inverse F-transform $f_{n_s}^F(t)$ based on the direct F-transform components $F_1^{(s)}, \ldots, F_h^{(s)}, \ldots, F_{n_s}^{(s)}$ is calculated, where

$$f_{n(s)}^F(t) = \sum_{h=1}^{n} F_h^{(s)} \cdot A_h^{(s)}(t) \tag{10.5}$$

The assessed value $\tilde{y}_0(t)$ of the measured output variable y_0 at the time t is given by the formula:

Fig. 10.5 Schema of the TSSF seasonal time series forecasting algorithm

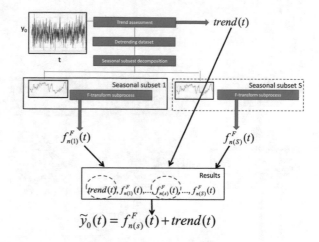

$$\tilde{y}_0(t) = f_{n(s)}^F(t) + trend(t) \tag{10.6}$$

where $trend(t)$ is the assessed value of the trend at the time t calculated via a polynomial best fit. The schema of this process is shown in Fig. 10.5.

In Di Martino and Sessa [3] a ninth-degree polynomial is applied to assess the trend function. The de-trended dataset is partitioned in S seasonal subset and initially the most coarse-grained fuzzy partition ($n_s = 3$) of any subset is created. If the coarse subset is sufficiently dense with respect to the fuzzy partition, the direct F-transform components and the MAD mean index are calculated. If then the MAD mean index is less or equal to a fixed threshold, the direct F-transform components are stored, otherwise a finer fuzzy partition ($n_s := n_s + 1$) is constructed. This process is repeated for any subset.

Below we schematize the flux diagram of the TSSF algorithm and the algorithm is shown in pseudocode (Fig. 10.6).

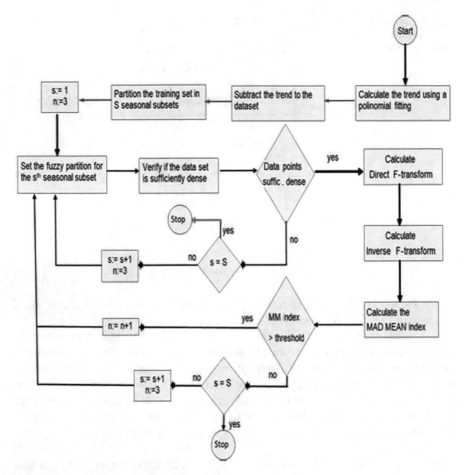

Fig. 10.6   Flux diagram of the TSSF algorithm

Algorithm: TSSF [3]	
Input:	*Seasonal time series* $y_1, y_2, ..., y_M$
	Threshold Mad mean value
Output:	*Polynomial fitting coefficients*
	Direct fuzzy transform components

1	Calculate the trend using a polynomial fitting
2	Store the polynomial fitting coefficients
3	*for* i = 1:M
4	Substract the trend to the measures obtaining $y(i) := y^0(i) - trend(t(i))$
5	*next* i
6	Partition the seasonal time series in S seasonal subsets
7	*for* s = 1:S
8	$n_s := 3$
9	stopIter := FALSE // stop cycle variable
10	*while* stopIter = FALSE
11	Set the uniform fuzzy partition
12	Verify the sufficiently density of the subset of data
13	*if* data are suff. dense with respect to the fuzzy partition *then*
14	Calculate the direct fuzzy transform components
15	Calculate the inverse fuzzy transform at any point
16	Calculate the Mad mean index
17	*if* Mad mean > threshold *then*
18	$n_s := n_s + 1$
19	*else*
20	Store the direct fuzzy transform components
21	stopIter := TRUE
22	*end if*
23	*else*
24	stopIter := TRUE
25	*end if*
26	*end while*
27	*next* s
28	*return*

Below in the pseudocode the function used to forecast the time series value at a time t *is shown.*

Algorithm: TSSF forecasting [3]	
Input:	*Time t*
	Polynomial fitting coefficients
	Seasonal direct F-transforms
Output:	*Forecasted value*
1	Select the season s in which the t period falls
2	Restore the direct F-transform components $F_1^{(s)}, ..., F_h^{(s)}, ..., F_{n_s}^{(s)}$
3	Calculate the inverse F-transform $f_{n(s)}^F(t)$ by (10.5)
4	Calculate *trend(t)* by using the polynomial fitting coefficients
5	Calculate $\tilde{y}_0(t)$ by (10.6)
6	*return $\tilde{y}_0(t)$*

The authors test the TSSF algorithm on a dataset daily weather data obtained by using measures collected every 30 min from the weather station Napoli Capodichino in the municipality of Naples (Italy).

A first experiment is performed by extracting the daily mean temperature in the months of July and August from 07/01/2003 till 08/16/2015. Figure 10.7 shows the trend obtained by using a ninth-degree polynomial fitting.

The MADMEAN threshold index is set to 6%. Comparison tests are performed with the Average seasonal variation method, Seasonal ARIMA method and Multidimensional F-transform method.

The graph plotted in Fig. 10.8 shows the value of mean temperature assessed by using the TDFF algorithm.

For comparing the four methods four indexes (RMSE, MAPE, MAD, and MAD MEAN) are calculated to evaluate the accuracy of the results, where

Fig. 10.7 Trend of the mean temperature obtained by using a ninth-degree polynomial fitting

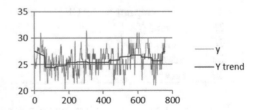

Fig. 10.8 Values of the mean temperature obtained by using the TSFF algorithm

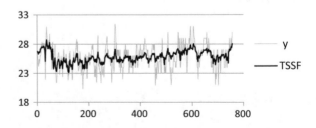

$$RMSE = \sqrt{\frac{\sum_{j=1}^{M} \left(\tilde{y}_0(t^{(j)}) - y_0^{(j)} \right)^2}{M}} \tag{10.7}$$

$$MAPE = \frac{1}{M} \sum_{j=1}^{M} \left| \frac{\tilde{y}_0(t^{(j)}) - y_0^{(j)}}{y_0^{(j)}} \right| \tag{10.8}$$

$$MAD = \frac{1}{M} \sum_{j=1}^{M} \left| \tilde{y}_0(t^{(j)}) - y_0^{(j)} \right| \tag{10.9}$$

$$MAD\,MEAN = \frac{\sum_{j=1}^{M} \left| \tilde{y}_0(t^{(j)}) - y_0^{(j)} \right|}{\sum_{j=1}^{M} y_0^{(j)}} \tag{10.10}$$

Table 10.1 shows the accuracy indexes calculated for the mean temperature time series by using the four algorithms.

The highest accuracy is obtained by using the TSSF algorithm. This result is confirmed also in other tests performed on dataset in which are used other weather parameters. Table 10.2 shows the RMSE value obtained by applying the four methods to test dataset for the parameters mean, min, and max temperature. The table reports the size of the training and of the test datasets, the season considered, and the RMSE calculated by applying the four methods.

The results in Table 10.2 show that the best forecasting performance in all the seasonal time series is obtained by using the TSFF algorithm.

Table 10.1   Accuracy indexes values obtained for the mean temperature time series [3]

Forecasting method	RMSE	MAPE (%)	MAD	MAD mean (%)
avg SV	1.78	5.60	1.42	5.50
ARIMA	1.55	4.89	1.24	4.80
F-transforms	1.61	4.96	1.28	5.05
TSSF	1.37	4.29	1.09	4.22

Table 10.2   RMSE index obtained on test datasets by using the four forecasting algorithms [3]

Parameter	Training dataset size	Test dataset size	Season	RMSE			
				Avg. SV	ARIMA	F-tr	TSSF
T. mean	757	15	Week	1.83	1.58	1.62	1.39
T. max	757	15	Week	2.20	1.92	1.99	1.68
T. min	4354	212	Month	2.90	1.27	1.54	0.94

10.4 F^1-Transform for Seasonal Time Series Forecasting

The inverse F^1-transform is a first-order refinement of the inverse F-transform for approximating a function: it can be used as a regression function to improve the performances obtained by using F-transform.

Di Martino and Sessa [4] propose a seasonal time series forecasting algorithm based on the inverse F^1 transform.

A polynomial fitting is applied to assess the trend of the time series, then the dataset $(t^{(1)}, y^{(1)}), (t^{(2)}, y^{(2)}) \ldots (t^{(M)}, y^{(M)})$ is detreated obtaining the time series given by

$$y^{(i)} = y_0^{(i)} - trend(t^{(i)}) \ i = 1, \ldots, M \tag{10.11}$$

This dataset is partitioned in S subsets, $\left\{ \left(t^{(1)}, y^{(1)}\right), \left(t^{(2)}, y^{(2)}\right) \ldots \left(t^{(M_s)}, y^{(M_s)}\right)\right\}$, $s = 1, 2, \ldots, S$, being s the seasonal period. Each subset represents the seasonal fluctuations with respect to the trend.

Let $\left[t_s^-, t_s^+\right]$ be the domain of sth subset and $\left\{A_1, A_2, \ldots, A_{n_s}\right\}$ be an h-uniform generalized fuzzy partition sufficiently dense with respect to this subset, where $A_1, A_2, \ldots, A_{n_s}$ are functions four times differentiable in the domain $\left[t_s^-, t_s^+\right]$.

The direct F^1-transform components are given by

$$F_k^1(t) = c_k^0 + c_k^1(t - t_k) \quad k = 1, \ldots, n_s \tag{10.12}$$

where

$$c_k^0 = \frac{\sum_{i=0}^{m_s} y^{(i)} A_k(t^{(i)})}{\sum_{i=0}^{m_s} A_k(t^{(i)})} \quad k = 1, \ldots, n_s \tag{10.13}$$

$$c_k^1 = \frac{\sum_{i=0}^{m_s} f(t^{(i)})(t^{(i)} - t_k) A_k(t^{(i)})}{\sum_{i=0}^{m_s} (t^{(i)} - t_k)^2 A_k(t^{(i)})} \quad k = 1, \ldots, n_s \tag{10.14}$$

In (10.10) c_k^0, $k = 1, \ldots, n_s$, are the components of the direct F^0-transform.

The seasonal fluctuations with respect to the trend are approximated by the inverse F^1-transform:

$$f_{n_s}^1(t) = \frac{\sum_{k=1}^{n_s} F_k^1(t) A_k(t)}{\sum_{k=1}^{n_s} A_k(t)} \tag{10.15}$$

To assess the value of y_0 at the sth season the following formula is applied:

$$\tilde{y}_0(t) = f_{n_s}^1(t) + trend(t) \tag{10.16}$$

In DI Martino and Sessa [4] the authors apply this algorithm, called Time Series Seasonal F^1-transform (for short, TSSF1) to weather seasonal time series. TSSF1

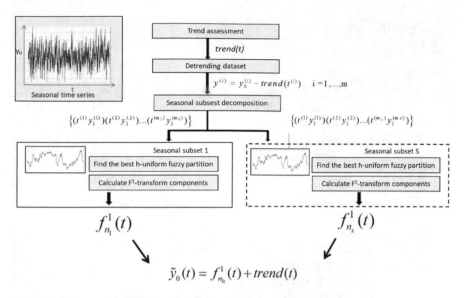

Fig. 10.9 Schema of the TSSF1 seasonal time series forecasting algorithm

is a variation of the TSSF algorithm in which the components of the direct F¹-transform (10.11) and (10.12) are calculated instead of the components of the direct F⁰-transform in order to improve the seasonal time series forecasting performances.

The schema of the TSSF1 algorithm is shown in Fig. 10.9.

Below in the pseudocode the TSSF1 algorithm is schematized.

Algorithm: TSSF1 [4]		
Input:		*Seasonal time series* $y_1, y_2,\ldots,y_M$
		Threshold Mad mean value
Output:		*Polynomial fitting coefficients*
		Components c_k^0 *and* c_k^1 $k = 1,\ldots, n_s$

1	Calculate the trend using a polynomial fitting
2	Store the polynomial fitting coefficients
3	*for* $i = 1{:}M$
4	Substract the trend to the measures obtaining $y(i) := y^0(i) -$ $trend(t(i))$
5	*next* i
6	Partition the seasonal time series in S seasonal subsets
7	*for* $s = 1{:}S$
8	$n_s := 3$
9	stopIter := FALSE // stop cycle variable
10	*while* stopIter = FALSE
11	Set the uniform fuzzy partition
12	Verify the sufficiently density of the subset of data
13	*if* data are suff. dense with respect to the fuzzy partition *then*
14	Calculate the components c_k^0 $k = 1, \ldots, n_s$ by (10.13)
15	Calculate the components c_k^1 $k = 1, \ldots, n_s$ by (10.14)
16	Calculate the direct fuzzy transform components at any point
17	Calculate the inverse fuzzy transform at any point
18	Calculate the MADMEAN index
19	*if* MADMEAN > threshold *then*
20	$n_s := n_s + 1$
21	*else*
22	Store the components c_k^0 and c_k^1 $k = 1, \ldots, n_s$
23	stopIter := TRUE
24	*end if*
25	*else*
26	stopIter := TRUE
27	*end if*
28	*end while*
29	*next* s
30	*return*

Below we show in pseudocode the function used to forecast the time series value at a time t occurred in the season s. The F^1-Transform components at the time t are reconstructed, then the F^1-transform $f_{n(s)}^1(t)$ at the time t is calculated.

Algorithm: TSSF forecasting [4]	
Input:	*Time t*
	Polynomial fitting coefficients
	Seasonal direct F-transforms
Output:	*Forecasted value*
1	Select the season s in which the t period falls
2	Restore the direct the components c_k^0 and c_k^1 $k = 1, ..., n_s$
3	Calculate the direct F-transform components $F_k^1(t)$ $k = 1, ..., n_s$ by (10.10)
4	Calculate the inverse F^1-transform $f_{n(s)}^1(t)$ by (10.15)
5	Calculate *trend(t)* by using the polynomial fitting coefficients
6	Calculate $\tilde{y}_0(t)$ by (10.16)
7	***return*** $\tilde{y}_0(t)$

In Di Martino and Sessa [4] the TSSF1 algorithm is applied on a set of time series composed of daily weather data collected from weather stations located in Italy in the Campania Region and managed by the Italian Air Force. The weather parameter analyzed in these time series is the Heat Index (HI), an index function of the max daily air temperature and of the daily relative humidity that measures the discomfort caused by the presence of heat waves in summer days. Each dataset contains the daily HI index measured in the months of July and August from July 1, 2003– August, 16 2017. The season is given by the week in the year. A ninth-degree polynomial fitting is applied to extract the trend function. In Fig. 10.10 the trend of the HI index in the time series is shown and obtained from the weather data measured form the station Napoli Capodichino.

Comparison tests of the TSSF1 algorithm are performed with the Seasonal ARIMA [1], ADANN [10, 11], and the TSSF algorithms. A GA algorithm with a stopping criterion of 200 generations to search the optimal number of the input and hidden layer nodes was applied in the ADANN method.

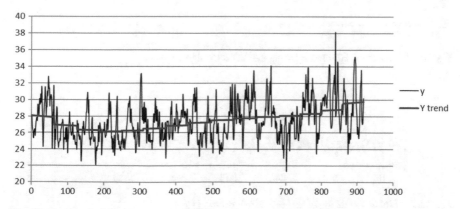

Fig. 10.10 Trend of the HI index obtained from the Napoli Capodichino station dataset [4]

In Figs. 10.11, 10.12, 10.13, and 10.14 the plot of the HI index obtained from the time series in Fig. 10.10 is shown by using the four methods.

In Table 10.3 the RMSE, MAPE, MAD, and MADMEAN accuracy indices are shown.

The results in Table 10.3 show that the better performances are obtained by using the ADANN and the TSFF1 methods, however, ADANN's computational complexity is higher than that of TSS1 in which a computational effort is spent to obtain the optimal number of input and hidden nodes.

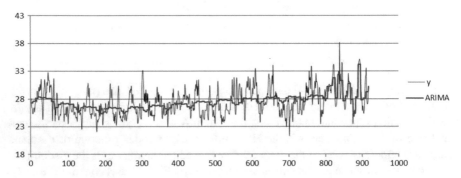

Fig. 10.11 Plot of the HI index by using the Seasonal ARIMA algorithm [4]

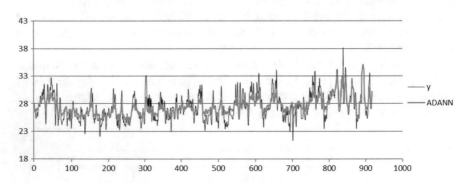

Fig. 10.12 Plot of the HI index by using the ADANN algorithm [4]

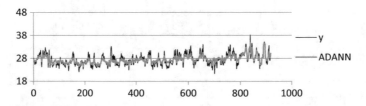

Fig. 10.13 Plot of the HI by using the TSSF algorithm [4]

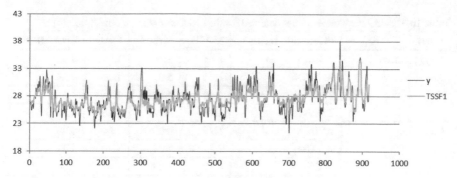

Fig. 10.14 Plot of the HI index by using the TSSF1 algorithm [4]

In the testing phase, a test dataset composed of the HI values measured in the period July 1, 2018–August 31, 2018 is used. The RMSE of the forecasted values with respect to the measured data is calculated. Table 10.4 shows the results obtained by using the four algorithms.

These results confirm that the best performances are obtained by using TSSF1 algorithm and ADANN. The RMSE, calculated by using the TSSF1 algorithm, is less than the one obtained by using the TSSF1 algorithm for all the datasets.

The results of these experiments show that multi-dimensional F-transform provides performance comparable to that one provided by the Wang and Mendel algorithm and by a Local Linear Wavelet Neural Network. Moreover, the former method has the advantage of verifying the density of the data with respect to the fuzzy partitions, determining which fuzzy partitions generate results that are performance acceptable.

The weak point of the multi-dimensional F-transform method is in the choice of the threshold values for the RMSE and MADMEAN indices: these thresholds can be determined by checking for which dimensionality of the fuzzy partitions the index value reaches a plateau where finer partitioning does not produce significant performance improvements.

Table 10.3 Accuracy indexes values obtained for the mean temperature time series [4]

Station	Forecasting method	RMSE	MAPE	MAD	MADMEAN
Capo Palinuro	ARIMA	1.65	5.56	1.54	4.95
	ADANN	1.43	5.22	1.24	4.38
	TSSF	1.49	5.37	1.34	4.56
	TSSF1	1.43	5.22	1.26	4.37
Capri	ARIMA	1.75	5.63	1.64	5.00
	ADANN	1.53	5.28	1.36	4.41
	TSSF	1.59	5.43	1.47	4.60
	TSSF1	1.52	5.30	1.37	4.41
Grazzanise	ARIMA	1.72	5.59	1.61	4.96
	ADANN	1.50	5.30	1.38	4.49
	TSSF	1.61	5.47	1.45	4.58
	TSSF1	1.53	5.29	1.36	4.45
Napoli Capodichino	ARIMA	1.68	5.48	1.41	4.93
	ADANN	1.46	5.14	1.17	4.35
	TSSF	1.52	5.29	1.26	4.54
	TSSF1	1.45	5.16	1.18	4.35
Salerno	ARIMA	1.74	5.63	1.61	4.98
	ADANN	1.52	5.34	1.38	4.51
	TSSF	1.63	5.51	1.45	4.60
	TSSF1	1.55	5.33	1.36	4.47
Pontecagnano	ARIMA	1.62	5.43	1.35	4.87
	ADANN	1.41	5.07	1.13	4.30
	TSSF	1.51	5.16	1.20	4.45
	TSSF1	1.39	5.06	1.13	4.29
Trevico	ARIMA	1.76	5.67	1.62	5.01
	ADANN	1.56	5.36	1.39	4.50
	TSSF	1.64	5.54	1.47	4.65
	TSSF1	1.55	5.36	1.38	4.51

Table 10.4 RMSE index obtained on test datasets by using the four forecasting algorithms [4]

Station	RMSE			
	ARIMA	ADANN	TSSF	TSSF1
Capo Palinuro	1.28	1.01	1.19	0.99
Capri	1.33	1.02	1.22	1.02
Grazzanise	1.35	1.04	1.24	1.05
Napoli Capodichino	1.35	1.04	1.22	1.03
Salerno	1.36	1.05	1.24	1.05
Pontecagnano	1.32	1.03	1.20	1.04

References

1. Wei, W. W. S. (2006). *Time series analysis univariate and multivariate methods* (2nd ed., 605 pp.). Pearson Addison Wesley. ISBN: 0-321-32216-9.
2. Novak, V., Perfilieva, I., Holcapek, M., & Kreinovich, V. (2014). Filtering out high frequencies in time series using F-transform. *Information Sciences, 274,* 192–209.
3. Di Martino, F., & Sessa, S. (2017). Time series seasonal analysis based on fuzzy transforms. *Symmetry, 9*(11), 281, 21 pp.
4. Di Martino, F., & Sessa, S (2019). Seasonal time series forecasting by F^1-fuzzy transform. *Sensors, 19*(16), 3611, 14 pp.
5. Cleveland, R., Cleveland, W., McRae, J. E., & Terpenning, I. (1990). STL: A seasonal-trend decomposition procedure based on Loess. *Journal of Official Statistics, 6,* 3–73.
6. Harvey, A., & Trimbur, T. (2003). General model-based filters for extracting cycles and trends in economic time series. *The Review of Economics and Statistics, 85,* 244–255.
7. Woodward, W. A., Gray, H. L., & Elliott, A. C. (2016). *Applied time series analysis with R* (2nd ed., 634 pp.). Boca Raton, FL: CRC Press. ISBN: 978-1498734226.
8. Hassani, H. (2007). Singular spectrum analysis: Methodology and comparison. *Journal of Data Science, 5,* 239–257.
9. Golyandina, N., & Zhigljavsky, A. (2013). *Singular spectrum analysis for time series*, Springer Briefs in Statistics. Berlin: Springer. ISBN: 978-3-642-34913-3.
10. Donate, J. P., Li, X., Sánchez, G. G., & Sanchis de Miguel, A. (2013). Time series forecasting by evolving artificial neural networks with genetic algorithms, differential evolution and estimation of distribution algorithm. *Neural Computing and Applications, 22,* 11–20.
11. Štepnicka, M., Cortez, P., Peralta Donate, J., & Štepnickova, L. (2013). Forecasting seasonal time series with computational intelligence: On recent methods and the potential of their combinations. *Expert Systems with Applications, 40,* 1981–1992.

Chapter 11
Fuzzy Transform for Data Classification

Data classification is a data mining supervised learning process aimed to classify a set of data points or patterns. As regression methods, the goal of classification algorithms is the construction of a model that can predict the value of the dependent variable from the independent ones. The difference between the regression and classification is the fact that the dependent attribute is numerical for regression and categorical for classification. Classes are sometimes called targets, labels, or categories.

In this chapter, the application of the multi-dimensional discrete F-transform for data classification is discussed. The multi-dimensional inverse F-transform can be applied in data classification similarly to how it is applied as a regression function to analyze dependency between numerical attributes in datasets. A classification method based on the multi-dimensional F-transform and applying the K-fold cross-validation method is analyzed in Sect. 11.5.

11.1 Machine Learning Data Classification: Underfitting and Overfitting

Machine learning studies how to automatically learn to acquire knowledge and make predictions based on past observations. In machine learning data classification is the problem of identifying to which of a set of classes a new observation belongs, starting from the knowledge acquired from the analysis of a training dataset composed of observations of which the class of belongingness is known. Each observation is given by a data point or pattern composed by measured values of the input variables which are also called independent variables or predictors. The output variable is also called the dependent variable and contains the label of the class assigned to the data point.

A classification model requires a training phase necessary to train the model to correctly classify the patterns and a testing phase in which the performances of the algorithm are measured by applying the classifier to a test set.

© Springer Nature Switzerland AG 2020
F. Di Martino and S. Sessa, *Fuzzy Transforms for Image Processing and Data Analysis*,
https://doi.org/10.1007/978-3-030-44613-0_11

The training set is composed of data point already correctly classified and is used to train the classifier. Machine learning classification encompasses many applications: examples of fields in which machine learning classification algorithms are applied are face detection, natural language processing, sentiment analysis, and bioinformatics.

The quality of a machine learning classifier is related to its capacity to be robust to underfitting and overfitting.

An underfitting problem occurs when the number of incorrectly classified patterns is not negligible: the evaluation of the presence of underfitting is given by measuring after the learning phase classification indexes as accuracy, recall, and precision.

Underfitting does not depend on the randomness of the training data but on the excessive simplification of the learning model, for example, if the number of nodes in a decision tree or of neurons in a network is insufficient. To solve the underfitting problem we must reconstruct the classification model using more information.

Overfitting occurs when, even if the data in the training set are optimally fitted, the classifier's performances measured using the test set are poor: in this case, the model adapts extremely well to the training set, providing very low training errors, but it fails to classify patterns belonging to the test set just as well. For example, in the training phase the number of neurons in the hidden layers of a neural network could be increased in order to obtain negligible training errors, but, nevertheless, the test error can be high: this is probably due to the presence of noisy data in the training sets to which the network has adapted during the training phase or to the fact that the training set is scanty.

There are many machine learning classification models in the literature. They can be divided into the following:

11.1.1   Logistic Regression

Logistic Regression [1] is a statistical method performing a predictive analysis to measure the relationship between the dependent variable and the independent variables (predictors), estimating probabilities through a logistic function: these probabilities are subsequently transformed into binary values and the result of the forecast is assigned to the most similar class.

A logistic regression algorithm uses a logistic function to model binary output variables, labeled, for example, as *true* or *false*. The logarithm of the probability that a point belongs to the class *true* is given by a linear combination of the input variables whose coefficients are calculated by using the maximum-likelihood method and by setting a threshold value for this probability the point is classified as belonging to the class *true* if this probability is greater than the threshold.

The logistic regression method can be extended to multi-values or continuous output variables.

This method is simple to implement and to use. In order to be applied, it needs that there are no high correlations among the predictors: this control can be assessed by calculating the correlation matrix among the input variables. The main restriction of this method is that it cannot be used in the presence of many input variables otherwise the model can result in overfitting,

11.1.2 Naive Bayes Classifier

Naïve Bayes [2, 3] is a classification method based on the Bayes theorem with an assumption of independence among input variables. A further constraint imposed in the method is that all input variables have the same impact on the result and no input variable determines the value of the output variable more than the others. The Bayes theorem is applied to determine the conditional probabilities of the characteristics of the problem assigning the point to the class with the highest probability value.

If the input variables are discrete, a multinomial probability distribution for the input variables is applied: in this case, the frequency histogram of the input variable is used, where the frequency of occurrence of a value in an input variable is given by the number of the data points in the training set in which this variable takes this value.

If the input variables are discrete, a Gaussian probability distribution is applied. Other naïve Bayes models in which are used continuous input variables discretize these variables and use Bernoulli probability distributions.

Despite its simplicity, naive Bayes can often outperform more sophisticated classification methods.

The main advantages of Naive Bayes algorithms are their robustness with respect to the presence of noise in the data and the fact that they are fast and easy to implement but their biggest disadvantage is the requirement of independency among the input variables In most of the real-life cases, they are dependent and this affects the performance of the classifier.

11.1.3 K-Nearest Neighbor

A K-Nearest-Neighbors (KNN) algorithm [4, 5] is a supervised classification algorithm in which a subset of labeled points is used to train to classify to label other points. To classify a new point, the algorithm analyzes the K labeled points closest to that new point, called its nearest neighbors, assigning to the new point the class to which the greatest number of nearby points belongs.

A distance measure must be used to calculate the distances between points: if numerical variables are used, common distance metrics are the Euclidean, Manhattan, and Minkowski distances. The Hamming distance is generally used in the presence of categorical variables.

The main advantage of the KNN algorithm is its simplicity of use and its ease of implementation. Furthermore, KNN does not require the insertion of specific parameters or the application of restrictions and constraints.

On the other hand, the computational complexity of the algorithm increases exponentially as the size of the training set and the number of input variables increase.

11.1.4 Decision Tree

Decision Tree ([6]—Chap. 18) builds classification or regression models in the form of a tree structure. Decision trees are built using a heuristic called recursive partitioning: the algorithm splits the dataset into smaller and smaller subsets developing, at the same time, an associated decision tree incrementally. The process stops when all the data points within the subsets are sufficiently homogenous, or another stopping criterion has been met.

The final structure is a tree with decision nodes and leaf nodes. A decision node has two or more branches and a leaf node represents a classification or decision. The topmost decision node in a tree which corresponds to the best predictor called root node. Decision trees can handle both categorical and numerical data.

The decision tree algorithm is simple to build and is computationally fast and it also provides good performance in terms of accuracy of classification results. However, it is subjected to the problem of overfitting: small variations of the training set can also determine variations that are consistent in the decision tree and complex decision trees can accurately classify the data points in the training set but fail in the classification of new data points.

11.1.5 Random Forest

Random Forests or Random Decision Forests [7–9] are an ensemble learning method in which a multitude of decision trees at training time is constructed. The class assigned is the class returned by the largest number of trees. The dataset is partitioned in more samples and for each sample is constructed a decision tree: the classification is obtained by averaging the results of many decision trees. This strategy allows to mitigate or eliminate the problem of overfitting. In fact, partitioning the training dataset into multiple subsets and building a decision tree for each subset, IT does not result in the error of accurately modeling a single training set.

Random Forest is more robust than the Decision Tree algorithm in avoiding overfitting the model, but it is more time-consuming than Decision Tree and additional computational resources are required to implement it.

11.1.6 *Artificial Neural Network*

An artificial neural network multilayer perceptron classifier consists of units called neurons, arranged in layers, which convert an input vector into an output, a categorical output in classification problems ([10, 11]—Chap. 6, [12]). The neural network is constructed from three types of layers: the input layer, whose nodes are associated with input variables, the hidden layers that are intermediate layers between the input and the output layer and place where all the computation is done and the output layer, associated with the output variable, that produces the classification result.

To the arc connecting two neurons belonging to a layer and the next layer is constructed is assigned a weight, respectively, initially set randomly. The neural network learns to classify patterns in the training phase in which the correct class for each pattern is known and the network trains by adjusting the weights to predict the correct class label of the sample patterns. An activation function defines if given neuron should be activated or not based on the weighted sum.

Various types of neural networks are proposed in the literature for data classification; they become very powerful when used in a hybrid system with many types of predictive neural networks.

After training the model, it needs to evaluate the performance of the classifier by using a test set. The accuracy of the model is measured calculating the percentage of misclassified data in the training and in the test set. The accuracy index is given by the ratio between the number of correctly classified data points and the total number of data points.

Underfitting is present if the accuracy measured in the training set is low. On the contrary, when the accuracy measured on the training set is much higher than the accuracy measured on the test set, there may be an overfitting problem. Hence, in a good classifier, the accuracy measured both in training and in the test set must be high.

Some techniques can be used to prevent the presence of overfitting. The following section deals with a more used method to prevent overfitting: the K-fold cross-validation method.

11.2 *K*-Folds Cross-Validation Techniques

The presence of underfitting can be corrected by increasing the complexity of the classifier model. The presence of overfitting, on the other hand, is more difficult to assess, some techniques can be used to prevent or mitigate overfitting.

A technique used in decision trees to prevent overfitting is pruning. There are two pruning strategies: pre-pruning and post-pruning. The first is to stop the construction of the tree before it adapts too much to the training set, post-pruning, on the other hand, involves a bottom-up approach that is executed once the construction of the model has been completed, the goal is to eliminate the superfluous parts of the tree.

Well-known techniques to prevent overfitting in machine learning classification are the resampling techniques, in which the training set is randomly partitioned in subsets and a subset is used to evaluate the classification accuracy.

Several resampling techniques are applied to evaluate the performance of the classifier and prevent overfitting, the most known method is called *cross-validation* method [13]. This technique uses a subset of the dataset to test the classifier; initially, some of the data is randomly removed from the dataset before training begins, then when training is done, the set of removed data is used to test the performance of classifier.

The simplest cross-validation method is the *holdout* method. In this method, the dataset is initially partitioned into two datasets, the former is used as a training set, the latter as a test set. The usual dimensions of the two subsets are 80% and 90% of the size of the initial dataset for the training set and 20% and 10% for the test set, respectively, but other partitions can also be used.

The advantage of this method is that it is usually preferable to the residual method and takes no longer to compute. However, its evaluation can have a high variance. The evaluation may depend heavily on which data points end up in the training set and which end up in the test set, and thus the evaluation may be significantly different depending on how the division is made.

The most popular resampling cross-validation technique is K-folds cross-validation technique. Turkey [14]: the training set is partitioned randomly in K subsets called folds. Initially, the first fold is used as test set (the validation set) and the other folds are merged to be used as training set. The algorithm is iterative: in any cycle, the kth fold, $k = 1, ..., K$, is the validation set applied to validate the classifier constructed in the training phase merging the other folds.

The mean of the K classifier results is used to classify a new pattern. The average of the K accuracy indices is calculated to evaluate the performances: it allows to train and to test the model K times on different subsets of the training dataset and build up a performance estimation of a machine learning model on unseen data.

In [15, 16] K-folds cross-validation is applied to evaluate and compare the performance of classification models. The advantage of K-folds cross-validation with respect to other resampling techniques is that in K-folds cross-validation each subset is considered once as validation set. In Fig. 11.1 a K-folds cross-validation method is schematized.

K-fold cross-validation method is more effective the larger the size of the source dataset; in fact, if it is applied to small datasets, its effect of reducing the overfitting in the data is poor; on the other hand, as the size of the dataset increases, the computational complexity of this method also increases.

Some other cross-validation techniques as *leave-one-out* cross-validation, in which K is given by the number of patterns N, provides a low bias, scilicet a low overestimation of the validation error, but a high variance between the n-folds, scilicet a poor accuracy. The leave-one-out technique is preferable when the number of patterns N is low.

Through an appropriate choice of the number of folds K, it is possible to obtain a trade-off between accuracy and reduction of overfitting: choosing a too low value

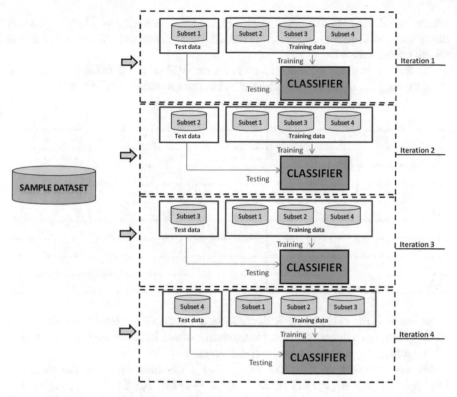

Fig. 11.1 Schema of a fourfold cross-validation technique [17]

for K, a high bias and a low variance would result. On the contrary, too high values of K produce a high variance and a low bias. $K = 5$ and 10 are the most common choices.

11.3   Multi-dimensional *F*-Transform for Data Classification

Let a training set be composed by a set of m instances (the patterns), characterized by a pair (X, Y), where X is a set of numerical attributes $(X_1, \ldots, X_s)$ and Y is a categorical attribute that takes values 1, ..., C, where $c = 1, \ldots, C$ denotes a class. The multi-dimensional F-transform can approximate the regression function as

$$Y = f(X_1, \ldots, X_s) \tag{11.1}$$

where f is a discrete function $f: [a_1, b_1] \times [a_2, b_2] \times \cdots \times [a_s, b_s] \rightarrow \{1, 2, \ldots, C\}$ and the real interval $[a_i, b_i]$ $i = 1, \ldots, s$ is the universe of discourse in which is defined the i*th* numerical attribute.

As for the analysis of dependency between attributes, the training set with m instances $O_1, \ldots, O_j, \ldots, O_m$ is schematized in the following tabular form:

	X_1	...	X_s	Y
O_1	p_{11}	.	p_{1s}	y_1
.	.	.	.	.
.	.	.	.	.
O_j	p_{j1}	.	p_{js}	y_j
.	.	.	.	.
.	.	.	.	.
.	.	.	.	.
O_m	p_{m1}	.	p_{ms}	y_m

where p_{ji} is the value of the attribute X_i in the instance O_j. Each attribute X_i can be considered as a numerical variable assuming values in the domain $[a_i, b_i]$, where $a_i = \min\{p_{1i}, \ldots, p_{mi}\}$ and $b_i = \max\{p_{1i}, \ldots, p_{mi}\}$.

We suppose that the function $f(x_1, \ldots, x_s)$ assumes values in the m points $(p_{j1}, p_{j2}, \ldots, p_{js}) \in [a_1, b_1] \times [a_2, b_2] \times \cdots \times [a_s, b_s]$ for $j = 1, \ldots, s$. Let $\{A_{11}, A_{12}, \ldots, A_{1n_1}\}, \ldots, \{A_{s1}, A_{s2}, \ldots, A_{sn_s}\}$ be a set of uniform fuzzy partitions. The set $P = \{(p_{11}, p_{12}, \ldots, p_{1s}), (p_{21}, p_{22}, \ldots, p_{2s}), \ldots, (p_{m1}, p_{m2}, \ldots, p_{ms})\}$ is sufficiently dense w.r.t. the chosen partitions $\{A_{11}, A_{12}, \ldots, A_{1n_1}\}, \ldots, \{A_{s1}, A_{s2}, \ldots, A_{sn_s}\}$ if for any $\{h_1, \ldots, h_s\} \in \{1, \ldots, n_1\} \times \cdots \times \{1, \ldots, n_s\}$ there is some $(p_{v1}, p_{h2}, \ldots, p_{vs})$ $\in P, v \in \{1, \ldots, m\}$, such that $A_{1h_1}(p_{v1}) \cdot A_{2h_2}(p_{v2}) \cdot \ldots \cdot A_{sh_s}(p_{vs}) > 0$. So we can define the $(h_1, h_2, \ldots, h_s)$th component $F_{h_1 h_2 \ldots h_s}$ of the direct F-transform of f w.r.t. the basic functions $\{A_{11}, A_{12}, \ldots, A_{1n_1}\}, \ldots, \{A_{s1}, A_{s2}, \ldots, A_{sn_s}\}$ as

$$F_{h_1 h_2 \ldots h_s} = \frac{\sum_{j=1}^{m} f(p_{j1}, p_{j2}, \ldots p_{js}) \cdot A_{1h_1}(p_{j1}) \cdot A_{2h_2}(p_{j2}) \cdot \ldots \cdot A_{sh_s}(p_{js})}{\sum_{j=1}^{m} A_{1h_1}(p_{j1}) \cdot A_{2h_2}(p_{j2}) \cdot \ldots \cdot A_{sh_s}(p_{js})}$$

(11.2)

The inverse F-transform of f w.r.t. the basic functions $\{A_{11}, A_{12}, \ldots, A_{1n_1}\}, \ldots, \{A_{s1}, A_{s2}, \ldots, A_{sn_s}\}$, to be the following function by setting for each point $p_j = (p_{j1}, p_{j2}, \ldots, p_{js}) \in [a_1, b_1] \times \cdots \times [a_s, b_s]$:

$$f_{n_1 n_2 \ldots n_s}^{F}(p_{j1}, p_{j2}, \ldots, p_{js}) = \sum_{h_1=1}^{n_1} \sum_{h_2=1}^{n_2} \cdots \sum_{h_s=1}^{n_s} F_{h_1 h_2 \ldots h_s} \cdot A_{1h_1}(p_{j1}) \cdot \ldots \cdot A_{sh_s}(p_{js})$$

(11.3)

for $j = 1, \ldots, s$. The jth pattern will be assigned to the class $\tilde{y}_j$ given by

$$\widehat{y}_j = \arg\left(\min_{c=1,\ldots,C} \left(\left| f^F_{n_1 n_2 \ldots n_s}(p_{j1}, p_{j2}, \ldots, p_{js}) - c \right| \right) \right) \qquad (11.4)$$

The percentage of misclassified instances in the training set is given by the percentage of instances for which $\widehat{y}_j \neq y_j$ ($j = 1, \ldots, m$.). It can be considered a training classification error.

11.4 K-Folds Cross-validation Applied to a Multi-dimensional F-Transform Classifier

The K-folds cross-validation method can be applied to a F-transform classifier in order to avoid overfitting problems.

The dataset is partitioned in K folds, at the kth iteration as validation set is used as the kth fold and as training set the set formed by merging the other folds. A set of uniform fuzzy partitions is created and the constraint of sufficient density of the data with respect to the fuzzy partition is verified.

The performances of the F-transform classifier can be measured by calculating a mean training error and a mean validation error where the mean is calculated by considering the errors obtained in any cycle.

Below in the pseudocode the F-transform classification algorithm is schematized.

K-folds cross validation F-transform classification	
Input:	*Training dataset*
	Number of folds K
Output:	*Direct F-transform components for the*
	K training set
	Mean training and validation errors

1	Partition randomly the dataset in K folds
2	*for* k = 1to K
3	Use the kth fold as validation set
4	Merge the other folds to form the kth training set
5	Set the uniform fuzzy partitions of the input variable domains
6	Verify the sufficiently density of the data points
7	Calculate the direct F-transform components as in (11.2)
8	Store the direct F-transform components
9	Classify each data point in the training set as in (11.4)
10	Calculate the training error
11	Store the training error
12	Classify each data point in the validation as in (11.6)
13	Calculate the validation error
14	Store the validation error
15	*next* k
16	Calculate a mean training error
17	Calculate a mean validation error
18	*return* mean training and validation errors
19	*end*

Let $P = (p_1, p_2, \ldots, p_s)$ be a point and $f^{F_k}_{n_1 n_2 \ldots n_s}(p_1, p_2, \ldots, p_s)$, $k = 1, \ldots, K$, be the value of the inverse F-transform obtained at the kth iteration. We can use a combination of the values of the inverse F-transforms to classify the point P. The simplest combination is the mean of the inverse fuzzy transforms, given by:

$$f_{n_1 n_2 \ldots n_s}(p_1, p_2, \ldots, p_s) = \frac{1}{K} \sum_{k=1}^{K} f^{F_k}_{n_1 n_2 \ldots n_s}(p_1, p_2, \ldots, p_s) \qquad (11.5)$$

The point P is classified via (11.4) obtaining that

$$\widehat{y} = \arg\left(\min_{c=1,\ldots,C}\left(\left| f_{n_1 n_2 \ldots n_s}(p_1, p_2, \ldots, p_s) - c \right| \right) \right) \qquad (11.6)$$

The main problem of this classification method is represented by the constraint of sufficient density that must be respected by each of the K training sets. A strategy to evaluate the optimal number of nodes of each fuzzy uniform partition similar to that experienced for the evaluation of the dependency between attributes must be implemented. The most coarse-grained partitions are built initially. After verifying that the constraint of sufficient data density with respect to the fuzzy partitions is respected and calculating the components of the multi-dimensional direct fuzzy transform, it is necessary to verify that the average error obtained in the classification of the data is less than a predetermined threshold. If this happens the algorithm terminates, otherwise it is necessary to consider finer fuzzy partitions and reiterate the process.

The *K*-folds cross validation *F*-transform classifier measures two type errors: training errors and validation errors. Following this strategy, to consider acceptable the classification, it needs to verify that both of these errors are below prefixed thresholds.

11.5   The MFC Algorithm

In [17] the authors propose a new data classification algorithm based on the multi-dimensional *F*-transform. They call this algorithm Multi-dimensional *F*-transform Classifier, for short MFC.

MFC uses an approach similar to the one applied in [18] to analyze dependency between attributes to find the optimal uniform fuzzy partitions of the domain of the input variables.

The dataset is partitioned in *K* folds and initially, a coarse-grained fuzzy partition of any input variable domain is created: if the data are sufficiently dense with respect to the fuzzy partitions, a training error CV_1 and a validation error CV_2 are calculated to measure the performances of the classifier. If CV_1 is less than a fixed training error threshold α and CV_1 is less than a fixed validation error threshold β, the process stops, else a set of finer fuzzy partitions of the domains of the input variables is considered and the process is iterated.

The two indices CV_1 and CV_2 are given by the mean training and validation errors, respectively, with respect to the errors calculated in any iteration:

$$CV_1 = \sum_{k=1}^{K} CV_1^k \tag{11.7}$$

$$CV_2 = \sum_{k=1}^{K} CV_2^k \tag{11.8}$$

where CV_1^k is the percentage of misclassified points in the *k*th training set and CV_2^k is the percentage of misclassified points in the *k*th validation set.

In Fig. 11.2 the MFC algorithm is schematized.

The components of the direct *F*-transforms are stored in any iteration and the mean of the inverse *F*-transforms (11.5) is calculated to classify a data point.

The algorithm is schematized below in pseudocode. The function Classify() is useful to assign to a class a point $P = (p_1, \ldots, p_s)$: it calculates the inverse *F*-transforms end return the identifier of the class to whom is assigned the point P. The algorithm

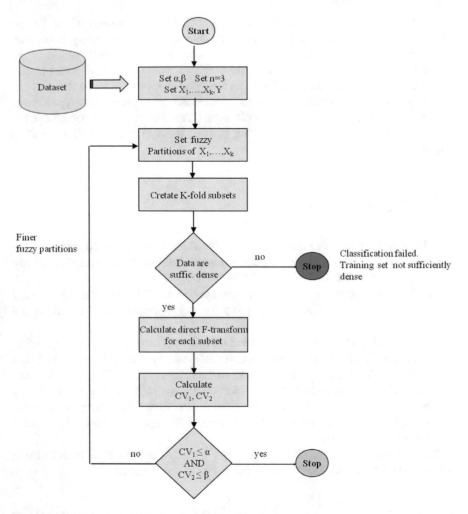

Fig. 11.2 Schema of the MFC algorithm [17]

calculates the percentage of misclassified patterns C_{1k} and C_{2k}, respectively, from the kth training and validation sets, in order to calculate the two errors C_1 and C_2. If both the two errors are below their respective thresholds α and β, then the Boolean variable stop Iteration is set to TRUE, the direct F-transform components are stored and the algorithm stops, else a finer set of uniform fuzzy partitions is created, and the process is iterated.

Algorithm: MFC (Di Martino & Sessa, 2020)	
Input:	*Dataset $(x_1^{(j)}, x_2^{(j)}, ..., x_k^{(j)}, y^{(j)}) j = 1, 2, ..., N$*
	Error thresholds α and β
Output:	*Direct Multiple F-transform components*

1	status:= FALSE
2	$n_1 := 3, ..., n_s := 3$
3	*while* (stopIteration == FALSE)
4	Partition randomly the dataset in K folds
5	*for* i = 1 to s
6	Create an uniform fuzzy partition $\{A_{i1}, A_{i2}, ..., A_{in}\}$ of the interval $[a_i, b_i]$ by using the basic functions (13)
7	*next* i
8	*for* k = 1 to K
9	*if* the data are not dense w.r.t. the fuzzy partitions *then*
10	stopIteration := TRUE
11	*return* -1 //*"Classification not found"*
12	*else*
13	Calculate the direct F-transform components by (11.2)
14	*end if*
15	*next* k
16	$C_1 := 0$ //training error
17	$C_2 := 0$ //validation error
18	*for* k = 1 to K
19	$C_{1k} := 0$ //training error k*th* training set
20	*for* each (p,y) in the k-th training set
21	y_f = Classify(C,P) // class assigned to the pattern p
22	*if* $(y_f \,!= y)$ *then*
23	$C_{1k} := C_{1k} + 1$
24	*end if*
25	$C_1 := C_1 + C_{1k}$
26	$C_{2k} := 0$ //validation error k*th* validation set
27	*for* each (p,y) in the k-th validation set
28	y_f = Classify(C,P) // class assigned to the pattern p
29	*if* $(y_f \,!= y)$ *then*
30	$C_{2k} := C_{2k} + 1$
31	*end if*
32	$C_2 := C_2 + C_{2k}$
33	*next* k
34	$C_1 := C_1 / K$

35	$C_2 := C_2 / K$
36	*if* $(C_1 > \alpha)$ OR $(C_2 > \beta)$ *then*
37	Set greater values for $n_1, ..., n_s$ // a finer fuzzy partitions set is selected.
38	*else*
39	stopIteration := TRUE
40	*for* k = 1 to K
41	Store the direct F-transform components
42	*next* k
43	*end if*
44	end while
45	*return* C_1, C_2 // *"Classification found"*
46	*end*

Algorithm: Classify	
Input:	Point P =$(p_1,...,p_s)$
	Number of classes C
	Direct F-transform components
Output:	*Class assigned to the point P*
1	$f: = 0$ // initialize to 0 mean inverse F-transform in the point P
2	$y: = 0$ // the point P is assigned initially to a null class
3	*for* k= 1 to K
4	Calculate the inverse F-Transform $f_{n_1 n_2 ... n_s}^{F^k}(p_1, p_2, ..., p_s)$ by (11.3)
5	$f := f + f_{n_1 n_2 ... n_s}^{F^k}(p_1, p_2, ..., p_s)$
6	*next* k
6	$f := f/K$
7	*for* c = 1 to C
8	*if* $|f - c| < 0.5$ *then*
9	y:= c
10	*end if*
11	*next* c
12	*return* y
13	*end*

A complex problem is to set appropriate values of the two parameters training error and validation error thresholds α and β: the lower these values are, the more accurate the classifier must be. However, by setting very low thresholds increase the risk that the MFC algorithm will fail due to the violation of the constraint of sufficient density with respect to fuzzy partitions too fine.

An acceptable solution is to archive the components of the multi-dimensional direct F-transform obtained using the finest possible fuzzy partitions in order to respect the constraint of sufficient density of all the training sets. In this case, the components of the direct F-transform will be updated in each cycle so that those stored at the end of the process will be the components of multi-dimensional direct the F-transform obtained using the finest fuzzy partitions for which the constraint of sufficient density is respected by all training sets. The rows 36–43 of the MFC algorithm will be changed to:

36	*for* k = 1 to K
37	Update the direct F-transform components
38	*next* k
39	*if* (C$_1$ > α) OR (C$_2$ > β) *then*
40	Set higher values for n$_1$, ..., n$_s$ // a finer fuzzy partitions set is selected.
41	*else*
42	stopIteration := TRUE
43	*end if*

In [17] the authors experiment with the use of the MFC algorithm by setting the parameters α and β, respectively, to 2 and 4% and the number of folds K to 10. They apply the MFC algorithm on over 100 datasets extracted from UCI machine Learning Knowledge Extraction Evolution Learning (KEEL) repositories. After the training phase, the classifier is tested using a testing dataset to measure the error rate, given by the percentage of misclassified instances. The following uniform fuzzy partition of the i input variable domain composed of cosinusoidal basic functions is used:

$$A_{i1}(x) = \begin{cases} 0.5 \cdot (1 + \cos \frac{\pi}{h_i}(x - x_{i1})) & \text{if } x \in [x_{i1}, x_{i2}] \\ 0 & \text{otherwise} \end{cases}$$

$$A_{ij}(x) = \begin{cases} 0.5 \cdot (1 + \cos \frac{\pi}{h_i}(x - x_{ij})) & \text{if } x \in [x_{i(j-1)}, x_{i(j+1)}] \\ 0 & \text{otherwise} \end{cases}$$

$$A_{in_i}(x) = \begin{cases} 0.5 \cdot (1 + \cos \frac{\pi}{h_i}(x - x_{in_i})) & \text{if } x \in [x_{i(n-1)}, x_{ini}] \\ 0 & \text{otherwise} \end{cases} \tag{11.9}$$

Below the results are shown for the IRIS dataset, which is formed by 150 instances and 5 attributes. The dataset consists of 150 samples, each of three species of plant: Iris Setosa, Iris Versicolour, and Iris Virginica.

The patterns are composed of the following attributes (Table 11.1).

The first four attributes constitute the input variables: they are the sepal and petal length and width of the flower measured in cm. The fifth attribute constitutes the output variable and is given by the type of Iris plant. It is well known that only the Iris Setosa is well linearly separable from the other two types of flower.

Each fold is composed of 15 data points, so each training set is composed of 135 points and each validation set of 15 points.

In Table 11.2 the value of the indices CV_1 and CV_2 and their averages are shown.

The resulting classifier is applied to a testing dataset of 50 instances downloaded from the IRIS WEB page project. The results in Table 11.3 show that MFC classifies

Table 11.1 Attributes in the IRIS dataset

Attribute	Description
Sepal length (cm)	Length of the sepal
Sepal width (cm)	Width of the sepal
Petal length (cm)	Length of the petal
Petal width (cm)	Width of the petal
Class	Type of iris plant 1—Iris Setosa 2—Iris Versicolour 3—Iris Virginica

Table 11.2 CV_1 and CV_2 from the IRIS dataset [17]

Fold	CV_1 (%)	CV_2
01	0.74	0.00
02	1.48	6.67
03	2.22	6.67
04	2.22	6.67
05	0.74	0.00
06	0.74	0.00
07	0.74	0.00
08	1.48	6.67
09	2.22	6.67
10	0.74	0.00
Mean	1.33	3.33

Table 11.3 Test results from the IRIS dataset [17]

Class	Number of instances	Wrongly classified instances	Test error rate (%)
Iris Setosa	17	0	0.00
Iris Versicolour	15	0	0.00
Iris Virginica	18	1	5.56
Total	50	1	2.00

improperly the data in the test dataset only in one case, the test error rate is 2%.

In [17] comparison tests are performed with other classification algorithms implemented in the mining tool WEKA (Waikato Environment for Knowledge Analysis, https://www.cs.waikato.ac.nz/ml/weka/). Comparisons are performed with Decision tree-based J48 [19, 20], Multilayer Perceptron [11, 12], Naive Bayes [2], Dimitoglou, [3], and Lazy K-Nearest Neighbor IBK [4, 5].

Table 11.4 shows the test error rates obtained by using the UCI machine learning

Table 11.4 Comparison results by using some datasets in the UCI machine learning repository [17]

Algorithm	Iris Setosa	Iris Setosa	Iris Setosa (%)
MFC Classsifier	IRIS	1	2.00
	Balance Scale	1	4.00
	Banana	3	1.00
Decision tree J48	IRIS	1	2.00
	Balance Scale	0	0.00
	Banana	2	0.67
Multilayer Perceptron	IRIS	1	2.00
	Balance Scale	0	0.00
	Banana	2	0.67
Naive Bayes	IRIS	2	4.00
	Balance Scale	1	4.00
	Banana	5	1.67
Lazy IBK	IRIS	2	4.00
	Balance Scale	1	4.00
	Banana	4	1.33

repository datasets Iris, Balance Scale, and Banana.

Table 11.5 shows the mean test error rate and the corresponding standard deviations obtained by applying the five classification algorithms to all the test sets.

These results show that the performance of the MFC algorithm, obtained by setting $\alpha = 2\%$ and $\beta = 4\%$, is comparable with those obtained by applying the Decision Tree J48 and the Multilayer Perceptron algorithms. Moreover, it produces better results than the Lazy IBK and Naive Bayes algorithms.

Further refinements could be applied in order to increase the performance of the MFC classifier. For example, in (11.5) instead of calculating the average of the inverse $K F$-transform values in the data point, a weighted average could be used by considering the weight inversely proportional to the training error.

Table 11.5 Comparison results in [17]

Algorithm	Mean test error rate (%)	Standard deviation (%)
MFC Classsifier	1.16	0.17
Decision tree J48	0.95	0.15
Multilayer Perceptron	0.98	0.16
Naive Bayes	3.21	0.41
Lazy IBK	1.38	0.25

Finally, further improvements to the model can be made measuring, in addition to accuracy, precision, and recall indices. These two indices provide additional information on the classifier's performance. Precision measures the ratio between the number of points correctly assigned to a class and the number of points assigned to that class. MFC could be improved adding further constraints to evaluate at each iteration also the average values of the precision and recall indices of the individual classes measured both for the training set and for the validation set, verifying that these values exceed predetermined thresholds. In this way, a wider set of restrictions on performance assessments can be assigned, even if doing so increases the number of parameters in the model.

References

1. Harrel, F. (2001). *Regression Modeling Strategies: With Applications to Linear Models, Logistic Regression, and Survival Analysis* (522 pp). New York: Springer. ISBN: 978-1-4419-2918-1.
2. Panda, M., & Matra, M. (2007). Network intrusion detection using Naive Bayes. *International Journal of Computer Science and Network Security, 12*, 258–263.
3. Dimitoglou, G., Adams, J.A., & Jim, C.M. (2013). Comparison of the C4.5 and a Naïve Bayes classifier for the prediction of lung cancer survivability. *International Journal of Computer Science and Applications, 6*(2), 256–261.
4. Aha, D.W. (Ed.) (1997). *Lazy Learning* (432 pp). Norwell (MA): Kluwer Academic Publishers. ISBN: 0792345843.
5. Jiang, L., Cai, Z., Wang, D., & Jiang, S. (2007). Survey of improving K-nearest-neighbor for Classification. In *Fourth International Conference on Fuzzy Systems and Knowledge Discovery* (5 pp). China: IEEE Computer Society Press. https://doi.org/10.1109/fskd.2007.552.
6. Shalev-Shwartz, S., & Ben-David, S. (2014). *Understanding Machine Learning* (449 pp). Cambridge University Press. ISBN: 978-1-107-05713-5.
7. Ho, T.K. (1995). Random decision forests. In *Proceedings of the 3rd International Conference on Document Analysis and Recognition* (pp 278–282). Montreal, QC, 14–16 August 1995.
8. Ho, T. K. (1998). The random subspace method for constructing decision forests. *IEEE Transactions on Pattern Analysis and Machine Intelligence, 20*(8), 832–844. https://doi.org/10.1109/34.709601.
9. Ho, T. K. (2002). A data complexity analysis of comparative advantages of decision forest constructors. *Pattern Analysis and Applications, 5*(2), 102–112.
10. Duda, R.O., Hart, P.E., Stork, D.G. (2001). *Pattern Classification* (2 ed., 617 pp). Wiley. ISBN 978-0471056690.
11. Pal, S.K., Mitra, S. (1992). Multilayer perceptron, fuzzy sets, and classification. *IEEE Transactions on Neural Networks, 3*(5), 683–697.
12. Chaudhuri, C. B., & Bhattacharya, U. (2007). Efficient training and improved performance of multilayer perceptron in pattern classification. *Neurocomputing, 34,* 11–27. https://doi.org/10.1016/S09252312(00)003052.
13. Kohavi, R. (1995). A study of cross-validation and bootstrap for accuracy estimation and model selection. In *Proceedings of the Fourteenth International Joint Conference on Artificial Intelligence* (Vol. 2, No. 12, pp. 1137–1143). San Mateo, CA: Morgan Kaufmann.
14. Tukey, J. (1958). Bias and confidence in not quite large samples. *Annals of Mathematical Science, 29,* 614.
15. Wong, T. T. (2015). Performance evaluation of classification algorithms by k-fold and leave-one-out cross validation. *Pattern Recognition, 48*(9), 2839–2846.

16. Wong, T. T. (2017). Parametric methods for comparing the performance of two classification algorithms evaluated by K-fold cross validation on multiple data sets. *Pattern Recognition, 65,* 97–107.
17. Di Martino, F., Sessa, S. (2020). A classification algorithm based on multi-dimensional fuzzy transforms. *Journal of Ambient Intelligence and Humanized Computing* (in press).
18. Di Martino, F., Loia, V., & Sessa, S. (2010). Fuzzy transforms method and attribute dependency in data analysis. *Information Sciences, 180*(4), 493–505.
19. Mitchell, T.M. (1997). *Machine Learning* (414 pp). Mc Graw-Hill, New York. ISBN: 9780070428072.
20. Bhargawa, N., Sharma, G., Bhargava, R., & Mathuria, M. (2013). Decision tree analysis on j48 algorithm for data mining. *International Journal of Advanced Research in Computer Science and Software Engineering, 3*(6), 1114–1119.

Chapter 12
Fuzzy Transform for Analyzing Massive Datasets

By massive data or big data, we mean a collection of data so extensive in terms of volume, speed of generation and acquisition and heterogeneity that it requires specific techniques and methods to be managed and explored. Massive data refers generally to datasets whose size and volume is so large that it exceeds the ability of traditional database systems to capture, store, manage and analyze them, and of traditional data mining and machine learning algorithms to extract and acquire knowledge.

Nowadays in many application fields, it needs to acquire and analyze massive data sources, for example, in sentiment analysis for exploring social data or in real-time systems in which sensor data are continuously received, stored, and analyzed.

In the last decade, algorithms were proposed to manage knowledge discovery processes applied to massive data. Examples are MapReduce-based methods applied to parallelize data management and data mining algorithms [1], and Hash-based machine learning algorithms applied to manage massive datasets [2].

Variations of F-transform techniques can be applied to analyze massive data. In this chapter, we concern with F-transform techniques applied in regression analysis for detecting relations between variables, in data classification and in spatiotemporal analysis.

12.1 Massive Data Definition and Concepts

Massive or big data is generally referred to as a dataset too large and complex to require frameworks and algorithms ad hoc to manage and extract information within a reasonable time. There is not a prefixed dimensional threshold beyond which a dataset becomes a massive dataset, since the continuous technological evolution determines the possibility of managing and analyzing data with algorithms whose calculation speeds are gradually increasing.

Specifically, a first characterization of massive datasets, known as 3Vs [3], is in Volume, Velocity, and Variety, where the term Volume is the size of the dataset, the term Velocity indicates the speed of data in and out, and the term Variety describes

© Springer Nature Switzerland AG 2020   193
F. Di Martino and S. Sessa, *Fuzzy Transforms for Image Processing and Data Analysis*,
https://doi.org/10.1007/978-3-030-44613-0_12

Table 12.1 Dataset size's classification [7]

Size (byte)	Class of dataset
10^6	Medium (M)
10^8	Large (L)
10^{10}	Huge
10^{12}	Monster
$>10^{12}$	Very Large (VL)

the variety and the inhomogeneity of data types and information. Sometimes, people extend another V according to their special requirements. Over time, other two V have been introduced: Veracity and Value, where the term Veracity indicates the accuracy, truthfulness, and reliability of the data and the term Value indicates the real value, in business terms, that will be produced by managing and analyzing the dataset [4, 5].

Additional characteristics of a massive dataset have recently been proposed [6]: one of them is the scalability, which refers to the rate of change in the time of the size of the dataset.

Massive data applications lie in many are as sentiment analysis and social mining, market analysis, sensor analytics, satellite big data analytics, web search engines, and recommender systems.

In the following, we will focus our interest on the dimensional characteristics of a massive dataset. A first classification of massive datasets based on dimensional characteristics was proposed in [7, 8]. Very Large and Large datasets [5] are produced and updated and they cannot be managed by traditional databases (Table 12.1).

For simplicity, in next paragraphs, we distinguish three types of datasets: Medium (M), Large (L), and Very Large (VL) datasets by considering as M dataset datasets sized less than 1 GB, as L a dataset taking up more than 1 GB in memory but less than 1 TB and as VL a dataset taking out more than 1 TB in memory. In practice, the sizes of a Large Dataset, even if extended, are such that they can be managed in memory by traditional algorithms. Following this definition, comparative experiments between machine learning and data mining algorithms implemented for massive datasets and traditional algorithms can be performed using appropriate L datasets.

In recent years, the use of emerging disciplines has been employed in the development of high-performance machine learning algorithms applied to massive datasets.

One of these disciplines is Quantum Computing Machine Learning [9–11] in which quantum algorithms are applied to improve classical machine learning techniques: such algorithms encode vector data as quantum information using qubits. The great advantage in the use of quantum computing lies in the fact that it is able to process massive data with speeds much higher than those of a traditional algorithm.

Another new field that represents a frontier of machine learning models is Deep Learning. Deng and Yu [12] based on deep neural networks, referred to artificial neural networks hierarchically structured in many layers, where each layer calculates the values for the next one so that the information is processed more and more completely. This architecture allows the network to be trained by using massive training sets: thanks to the introduction of Graphic Processing Units (GPUs), the training speeds are today much higher than in the past, and a deep neural network can process massive data in acceptable times providing extremely accurate results.

In this chapter, we will explore variations of machine learning techniques based on the use of multi-dimensional F-transform for the management of massive data. The aim of these techniques is to be usable in the presence of massive training sets while guaranteeing acceptable performances in terms of accuracy of results.

In the following sections, methods based on the use of multi-dimensional F-transform applied as a regression and classification function to massive datasets will be discussed. In order to compare the results with those obtained by applying traditional F-transform based algorithms, the experiments are performed on different L-datasets.

12.2 Massive Data Regression Analysis

Massive or big data is generally referred to as a dataset too large and complex to require frameworks and algorithms ad hoc to manage and to extract.

Let $X_1, \ldots, X_i, \ldots, X_r$ be r input variables. In a linear regression model, the output variable is considered as linear combination of the input variables, given by the formula:

$$Y = \beta_0 + \beta_1 X_1 + \beta_2 X_2 + \cdots + \beta_r X_r + \varepsilon \tag{12.1}$$

where the coefficients $_0, _1, \ldots,$ are the regression parameters, and is the error of the model. A regression machine learning model is applied to evaluate the regression parameters $'0, '1, \ldots, '$ by using a training set. The regression function Y in (12.1) is approximated as follows.

$$Y' = \beta_0' + \beta_1' X_1 + \beta_2' X_2 + \cdots + \beta_r' X_r \tag{12.2}$$

A traditional regression machine learning model cannot be applied if the training set is a massive dataset. A technique could be adopted to reduce the size of the training set. A simple method is to extract a random medium size subset by using it as training set. The main flaw of this method consists in the fact that this reduction of the size of the training set could seriously affect the accuracy of the regression model.

A random sampling strategy can be applied to increase the accuracy of the model. Various sampling techniques were proposed for big data analysis [13, 14].

In [15], a random sampling strategy is applied in massive data regression. In a random sampling, all data points in the dataset have equal chances to be selected to sample. The dataset is randomly partitioned in s subsets with similar size where each subset must constitute a representative sample of the population. The output variable is assessed by considering a combination of the output values as

$$Y' = f(Y'_{(1)}, Y'_{(2)}, \ldots, Y'_{(r)}) \tag{12.3}$$

In [15], the function f is given by the average of the output values assessed by applying the regression model to each subset. The regression parameters obtained are given by the averages of the regression parameters calculated for each subset:

$$\beta_k = \frac{1}{s} \sum_{l=1}^{s} \beta_{k,l} \quad k = 0, \ldots, r \tag{12.4}$$

To measure the accuracy of the results the Mean Square Error (MSE) in any subset given by:

$$MSE_l = \frac{1}{n_l} \sum_{i=1}^{n_l} \left(Y'(x_{1i}, x_{2i}, \ldots, x_{ri}) - y_i \right)^2 \quad l = 1, \ldots, s \tag{12.5}$$

is calculated, where $(x_{1i}, x_{2i}, \ldots, x_{ri}, y_i)$ are the input values and the output value of the ith data point in the lth subset and n_l is the size of the lth subset.

This method is schematized below:

Algorithm: Random sampling regression (Jun et al., 2015)	
Input:	Dataset of points P = $(x_{1i}, x_{2i}, \ldots, x_{ri}, y_i)$
	Number of subsets s
Output:	*Mean regression parameters*
	Mean MSE
1	Partition randomly the dataset in s equally sized subsets
2	*for* $l = 1$ to s
3	Apply the regression model to calculate the regression parameters
4	*for* i= 1 to n_l
5	Calculate y'_i by (12.3)
6	$MSE_l := MSE_l + (y'_I - y'_i)^2$
7	*next* i
6	$MSE_l := MSE_l / n_l$
	next l
8	Store the mean regression parameters $\beta_0, \beta_1, \ldots, \beta_r$
9	Calculate the mean MSE
10	*return* the mean MSE
11	*end*

The authors test this method with the Bike sharing dataset in the UCI machine repository: this dataset is composed of 16 attributes and over 17,000 data points. The aim of the test is to analyze the influence of temperature and humidity on bike rental:

Table 12.2 Comparison results in [15]

Fold	β_0	β_1	β_2	MSE
Random sampling	184.0596	361.9387	−278.1910	24,659.18
Traditional	184.2446	361.8051	−278.3578	24,643.72

the output variable is the number of total rental bikes and the two input variables are the temperature and the humidity. The dataset is partitioned in ten subsets using a simple random sampling without replacement and the regression method is applied to any subset. To evaluate the accuracy of this method, the authors apply traditionally the regression model to the overall dataset in order to compare the results with the one obtained using a random sampling technique. These results are shown in Table 12.2.

These results show that the accuracy obtained using the random sampling regression model is comparable with the one obtained by using the traditional model. In particular, the regression parameters and the MSE index obtained by partitioning the dataset in 10 subsets are similar to the ones obtaining applying the traditional regression method on the overall dataset.

12.2.1 Fuzzy Transform for Analyzing Attribute Dependency in Massive Datasets

The multi-dimensional inverse F-transform was applied in [16, 17] to analyze attribute dependency in dataset. In Chap. 9 the method of using multi-dimensional F-transform as a regression function for the analysis of the dependency between attributes in datasets was explored in detail.

In particular, in [17] the authors adopt a strategy to obtain a trade-off between the fulfillment of the constraint of sufficient density of the data with respect to the fuzzy partitions and the accuracy of the regression model. This method, called F-transform Attribute Dependences (for short, FAD), is given by an iterative process in which the components of the multi-dimensional direct F-transform are stored. This process is summarized below:

1. Initially the smallest number of nodes is fixed ($n = 3$) is set;
2. the basic functions of the domains of the input variables are constructed;
3. if the data in the training set are not sufficiently dense with respect to the fuzzy partitions, the process stops and the dependency of the output variable from the input variables is not found;
4. the components of the multi-dimensional direct fuzzy transform are calculated by using the training data;
5. the multi-dimensional inverse F-transform for the training data is calculated and the index of determinacy is calculated in order to evaluate the accuracy of the model;

6. if the index of determinacy is greater than a fixed threshold the process stops, the dependency of the output variable from the input variables is found and the components of the multi-dimensional direct transforms are stored, else the number of nodes n is increased by 1 and the process goes back to step 2.

The main advantages of the FAD algorithm are its high usability and a good regression accuracy; however, it cannot be applied for massive data due to the impossibility of managing a very large training set in memory. The FAD algorithm, in fact, is memory resident as its use requires to handle the data and processes in memory.

Some authors proposed variations of nonlinear machine learning regression models to manage massive data. In [18, 19] a fast-local Support Vector Machine (SVM) method to manage L datasets is presented in which a set of multiple local SVMs for low-dimensional data are constructed. Zheng et al. [20] propose an incremental version of the vector machine regression model to manage large-scale data. Peng et al. [21] proposed a parallel architecture of a logistic regression model for massive data management. Recently variations of the Extreme Learning Machine (ELM) regression methods [22] based on MapReduce framework and applied to massive data are proposed [23–25].

These parallel MapReduce-based ELM variations can handle massive dataset with good performances in terms of accuracy and learning speed, however, similarly to SVM and Multilayer Perceptron for massive data algorithms, they are difficult to implement due to the presence of a high number of parameters.

An extension of the FAD algorithm is proposed in [26] to handle massive datasets, maintaining the accuracy performances and the usability of the FAD.

The authors apply the random sampling technique described in [15], subdividing the dataset into s subsets of equal or similar cardinality.

Formally, we consider a massive dataset with m data points schematized in tabular form as

	X_1	...	X_i	...	s
O_1	p_{11}	.	p_{1i}	.	p_{1r}
.	.	.	.	.	.
.	.	.	.	.	.
.	.	.	.	.	.
O_j	p_{j1}	.	p_{ji}	.	p_{jr}
.	.	.	.	.	.
.	.	.	.	.	.
.	.	.	.	.	.
O_m	p_{m1}	.	p_{mi}	.	p_{mr}

where $X_1,..., X_i, ..., X_r$ are the involved attributes, $O_1, ..., O_j, ..., O_m$ $(m > r)$ are the data points or instances and p_{ji} is the value of the attribute X_i in the data point O_j. Each attribute X_i is considered as numerical variable assuming values in the domain $[a_i, b_i]$, where $a_i = \min\{p_{1i},\dots, p_{mi}\}$ and $b_i = \max\{p_{1i},\dots, p_{mi}\}$. The Massive FAD algorithm (for short, MFAD) uses a massive dataset to evaluate the functional dependency between attributes in the form:

$$X_z = H(X_1,\dots, X_k) \tag{12.6}$$

where $z \in \{1,\dots,r\}, k \le r < m, X_z \ne X_1, X_2,\dots, X_k,, H: [a_1, b_1] \times [a_2, b_2] \times \cdots \times [a_k, b_k] \to [a_z, b_z]$.

After partitioning randomly, the dataset in s subsets, the FAD algorithm is applied independently to all subsets.

Let $D_l = [a_{1l}, b_{1l}] \times \cdots \times [a_{1l}, b_{1l}]$ with $l = 1,\dots,$ s be the Cartesian product of the domains of the attributes $X_1, X_2,\dots, X_k$, where a_{il} and b_{il} are the minimum and maximum values of X_i in the lth subset. A uniform fuzzy partition of n_i basic functions $\{A_{i1},\dots, A_{ij},\dots, A_{in_i}\}$ of the domain of all the input variable $[a_{il}, b_{il}]$, $i = 1, 2,\dots, k$, is created.

Let $P_j = (p_{jl}, p_{j2},\dots, p_{jk}, p_{jz})$ be the jth data point in the lth subset. If the data are sufficiently dense with respect to the fuzzy partitions, we can calculate the direct fuzzy transform $\mathbf{F}_l$ with components:

$$F_{h_1 h_2 \dots h_k} = \frac{\sum_{j=1}^m p_{jz} \cdot A_{1h_1}(p_{j1}) \cdot \dots \cdot A_{kh_K}(p_{jk})}{\sum_{j=1}^m A_{1h_1}(p_{j1}) \cdot \dots \cdot A_{kh_K}(p_{jk})} \tag{12.7}$$

Hence the multi-dimensional inverse F-transform $H^F_{n_1 n_{2l} \dots n_{kl}}$ is calculated for approximating the function H in the domain D_l. The multi-dimensional inverse F-transform in the point P_j is given by:

$$H^F_{n_{1l} n_{2l} \dots n_{kl}}(p_{j1}, p_{j2}, \dots p_{jk}) = \sum_{h_1=1}^{n_{1l}} \sum_{h_2=1}^{n_{2l}} \dots \sum_{h_k=1}^{n_{kl}} F_{h_1 h_2 \dots h_K} \cdot A_{1h_1}(p_{j1}) \cdot \dots \cdot A_{kh_K}(p_{jk}) \tag{12.8}$$

The accuracy is evaluated by measuring the index of determinacy given by

$$r_c^2 = \frac{\sum_{j=1}^m \left(H^F_{n_1 n_2 \dots n_k}(p_{j1}, p_{j2}, \dots p_{jk}) - \hat{p}_z\right)^2}{\sum_{j=1}^m \left(p_{jz} - \hat{p}_z\right)^2} \tag{12.9}$$

where $\hat{p}_z$ is the mean value of X_z in the lth subset. In MFAD the following variation of (12.9) is used and applied in the FAD algorithm in order consider the size of the subset m and the number of input variables k:

$$r_c'^2 = 1 - \left[(1 - r_c^2) \cdot \frac{m-1}{m-k-1}\right] \tag{12.10}$$

Let $P = (x_1, x_2, \ldots, x_k)$ be a point of $\bigcup_{l=1}^{s} D_l$. The MFAD algorithm approximates the value x_z with the weighted average of the multi-dimensional inverse F-transform calculated from all the s subsets in the point P. The weight w_l assigned to $H^F_{n_{1l}n_{2l}\ldots n_{kl}}$ is given by the modified index of determinacy (12.10) if $P \in D_l$. If the point $P \in D_l$, then the weight is zero. Then the formula for the weight w_l is given by

$$w_l(x_1, x_2, \ldots, x_k) = \begin{cases} r'^2_{cl} & \text{if } (x_1, x_2, \ldots, x_k) \in D_l \\ 0 & \text{otherwise} \end{cases} \quad l = 1, \ldots, s \qquad (12.11)$$

The value x_z is approximated by

$$H^F(x_1, x_2, \ldots, x_k) = \frac{\sum_{l=1}^{s} w_l(x_1, x_2, \ldots, x_k) \cdot H^F_{n_l}(x_1, x_2, \ldots, x_k)}{\sum_{l=1}^{s} w_l(x_1, x_2, \ldots, x_k)} \qquad (12.12)$$

where, for brevity, $H^F_{n_l}(x_1, x_2, \ldots, x_k) = H^F_{n_{1l}n_{2l}\ldots n_{kl}}(x_1, x_2, \ldots, x_k)$. The reason that the weights are given by the indices of determinacy obtained by applying the FAD algorithm to each subset is that the greater the value of the index of determinacy, the greater the accuracy of the results is.

In general, the FAD algorithm applied on two distinct subsets generates different fuzzy partitions and different values of the determinacy index. The index of determinacy, which estimates the accuracy of the prediction, therefore represents the most suitable index to be used as weight (12.10). The weight w_l is equal to 0 if the point P does not belong to the domain D_l, and if the lth subset is not sufficiently dense even with respect to the most coarse-grained fuzzy partitions ($n_{1l} = n_{2l} = \cdots = n_{kl} = 3$) and the components of the direct F-transform cannot be calculated by the FAD algorithm. The architecture of the MFAD algorithm is schematized in Fig. 12.1.

Initially, the massive dataset is randomly partitioned into s subsets of approximately equal size. MFAD executes the FAD algorithm in each cycle by storing the components of the direct F-transform $\mathbf{F}_l$, the calculated reliability index r'^2_{cl} and the domain D_l. Below we show in pseudocode the MFAD algorithm.

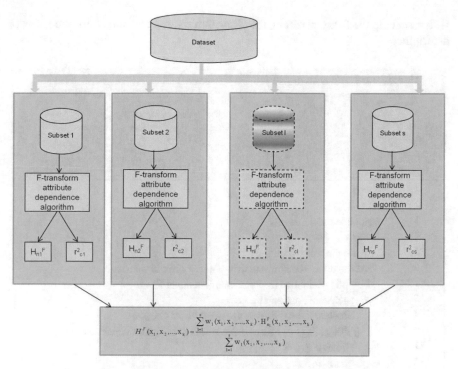

Fig. 12.1 Schema of the architecture of the MFAD algorithm

Algorithm: MFAD (Di Martino & Sessa, 2020a)	
Input:	Massive dataset DT composed by m instances with attributes $X_1, X_2,..., X_k, X_z$
	Number of subsets s
Output:	Direct F-transform components found for each subset
	Domains $D_1,..., D_s$
	Indexes of determinacy $r'^2_{c1},..., r'^2_{cs}$
1	Partition randomly the dataset DT in s subsets $DT_1,..., DT_s$
2	*for* $l = 1$ to s
3	$r'^2_{cl} := 0$
4	Calculate $\mathbf{F}_l, r'^2_{cl}, D_l$ via FAD algorithm
5	Store $\mathbf{F}_l, r'^2_{cl}, D_l$
6	*next l*
7	*end*

Input of the FAD algorithm is the training subset and the threshold of the index of determinacy α. The algorithm returns the direct F-transform matrix, the index of determinacy calculated and the domain D. If the training set is not sufficiently dense with respect to the fuzzy partition, the algorithm returns the value 0 for the index of determinacy.

Algorithm: FAD (Di Martino & Sessa, 2010a)	
Input:	Massive dataset DT composed by m instances with attributes $X_1, X_2,..., X_k, X_z$ threshold value α
Output:	Index of determinacy r'^2_c
	Direct F-transform matrix F
	Domain D

1	n := 3
2	$r'^2_c := 0$
3	Calculate the domain D
4	*while* $r'^2_c \leq \alpha$
5	Create the fuzzy partitions
6	Set the direct transform matrix **F** as a matrix of matrix n^k
7	F := [-1] //initialize to -1 the components of F
8	*for* each combination $\{h_1,..,h_k\}$
9	Calculate the F-Transform component $F_{h_1 h_2...h_k}$
10	*if* F[$h_1,...,h_k$] := -1 then // the training set is not sufficiently dense
11	*return* **F**, 0, D
12	*next* $\{h_1,...,h_k\}$
13	*else*
14	Calculate the index of determinacy r'^2_c
15	n:=n+1
16	*end while*
17	*return* **F**, r'^2_c, D
18	*end*

The algorithm *MFADEvaluation* returns the assessed value for the attribute Xz, given a point P with input coordinates $(x_1, x_2, ..., x_k)$. It uses (12.11) to calculate the weights and (12.12) to calculate to assess the value x_z.

The algorithm returns the assessed value. If all the weights are null, an error message is returned.

Algorithm: MFADEvaluation	
Input:	Point P = $(x_1, x_2, \ldots, x_k)$
	Direct F-trasforms $\mathbf{F_1}$ $\mathbf{F_2,}\ldots,$ $\mathbf{Fs}$
	Indices of determinacy $r'^2_{c1}, r'^2_{c2}, \ldots, r'^2_{cs}$
	Domains $D_1, \ldots, D_s$
Output:	Assessed value x_z
1	Calculate $H^F(x_1, x_2, \ldots, x_k)$ by (12.10)
2	w := 0
3	$wH_l := 0$
4	*for* $l = 1$ to s
5	*if* $(P \in D_l.)$ *then*
6	$w_l := r'^2_{cl}$
7	*else*
8	$w_l := 0$
	end if
9	w := w + w_l
10	$wH_l := wH_{l+}w_l$ * $H^F_{n_l}(x_1, x_2, \ldots, x_k)$
11	*next l*
12	*if* (w > 0) *then*
13	*return* wH_l / w
14	*else*
15	*return ERROR*
16	*end*

The accuracy of MFAD was measured in [26], applying the algorithm on a large dataset given by the census data of 402,678 census tracts of the Italian regions provided by the Italian National Statistical Institute (ISTAT) in 2011.

The dataset is composed of 140 numerical attributes belonging to the following categories:

- inhabitants,
- foreigner and stateless inhabitants,
- families,
- buildings,
- dwellings.

The FAD method is applied to the overall dataset, the MFAD method is applied by partitioning the dataset in s subsets and we perform the tests varying the value of the parameter s and by setting the threshold $\alpha = 0.7$.

In addition, we compare the MFAD algorithm with the Support Vector Regression (SVR) and Multilayer Perceptron (MP) algorithms. In this experiment, we explore the relation between the density in number of inhabitants per square kilometer of residents with a university degree and the density of resident population employed in order to verify if a higher density of population with a university degree should correspond a greater density of population employed. The attribute dependency explored is $H_z = H(X_1)$, where

Table 12.3 Mean index of determinacy for various values of s [26]

s	Index of determinacy
1	0.760
8	0.745
9	0.748
10	0.750
11	0.752
13	0.754
16	0.758
20	0.752
26	0.748
40	0.744

Input attribute: X_1 = Density of residents with university degree
Output attribute: Xz = Density of residents over 15 employed

Numerous tests have been carried out applying the MFAD algorithm for different values of the number of subsets s. In Table 12.3, we show the mean value of the index of determinacy obtained for different values of s. In the first test ($s = 1$) the FAD algorithm is applied to the whole large dataset.

The results in Table 12.3 show that the accuracy of the MFAD method is comparable with the one obtained by applying the traditional FAD method to the large dataset ($s = 1$, index of determinacy $= 0.760$). The best mean index of determinacy measured with the MFAD algorithm (0.758) is obtained for $s = 16$. Figure 12.2 shows the trend of $H_F(x_1)$ obtained for various values of s.

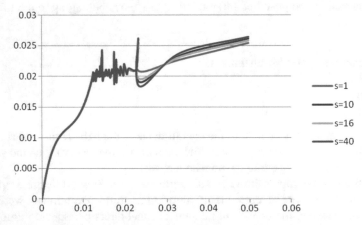

Fig. 12.2 Trends of $H_F(x_1)$ with respect to FAD method ($s = 1$)

Table 12.4 Mean index of determinacy for various values of s [26]

s	Index of determinacy
1	0.785
8	0.776
9	0.776
10	0.778
11	0.780
13	0.781
16	0.781
20	0.780
26	0.779
40	0.777

In another experiment, the dependency of the attribute with respect to the attributes density of residential buildings built with reinforced concrete and density of residential buildings built after 2005 is explored. The MFAD algorithm is applied to explore the dependency: $X_z = H_F(X_1, X_2)$, with input variables:

X_1 = Density of residential buildings built with reinforced concrete
X_2 = Density of residential buildings built after 2005

and output variable:

X_z = Density of residential buildings with state of good conservation

The threshold value α is set to 0.75. In Table 12.4 we show the mean indices of determinacy obtained for different values of the number s of subsets.

The results in Table 12.4 confirm that the accuracy obtained by applying the MFAD method is comparable with that one obtained with the traditional FAD method to the L dataset ($s = 1$, index of determinacy $= 0.785$). The best values are obtained with $s = 13$ and 16 (index of determinacy $= 0.781$).

The authors perform an analysis of the CPU time in this experiment measuring the gain with respect to the CPU time spent by applying the FAD algorithm to the whole dataset ($s = 1$) t_1. If t_s is the MFAD CPU time measured partitioning the dataset in s subset, the CPU time gain is given by the formula

$$t_s = \frac{t_1 - t_s}{t_1} \tag{12.13}$$

The trend of the CPU time obtained is shown in Fig. 12.3.

Th CPU time gain is null for $s = 1$ and is always positive for $s > 1$ and the greatest CPU time gain value is obtained for $s = 16$.

In this experiment, the authors compare the performances of MFAD with the ones obtained by applying the Support Vector Regression (SVR) and Multilayer Perceptron (MP) algorithms.

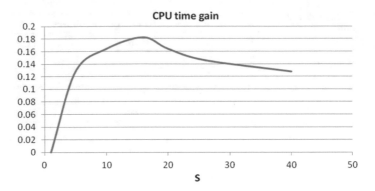

Fig. 12.3 Trend of CPU time gain with respect to the number of subsets s [26]

SVR is used in various kernel functions (linear, polynomial, Pearson VII universal kernel, and Radial Basin), and varying the complexity parameter C between 0 and 10. To compare the performances of the SVR and MFAD algorithms, the index of determinacy (12.10) is measured in each experiment. experiment.

Figure 12.4 shows the trend of the difference between the index of determinacy obtained in each test using the SVR algorithm and the mean index of determinacy obtained by using the MFAD algorithm, with respect to the index of the determinacy obtained using the SVR algorithm.

This difference is always under 0.002: then it is possible to conclude that MFAD produces accuracy results comparable with the ones obtained using the SVR algorithm, regardless of the function kernel and the C parameter chosen.

Similar tests are executed by comparing the performances of MFAD and MP, in which different values are used for the learning rate and momentum parameters in the range [0.1, 1] and one hidden layer is used in which the number of nodes varies between 2 and 8, the number of epochs is set to 500 and the percentage size of validation to 0.

Fig. 12.4 Trend of the difference between the index of determinacy obtained in SVR and MFAD

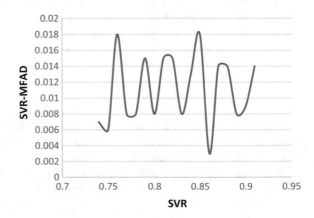

Fig. 12.5 Trend of the
difference between the index
of determinacy obtained in
MP and MFAD

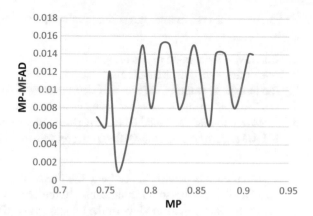

Figure 12.5 shows the trend of the difference between the index of determinacy obtained in each test using the MP algorithm and the mean index of determinacy obtained by using the MFAD algorithm, with respect to the index of the determinacy obtained using the MP algorithm.

Figure 12.5 shows that MFAD produces mean accuracies comparable even with the ones obtained using the MP algorithm, regardless of the values of the parameters set using MP that the index of determinacy difference is always under 0.16.

Therefore MFAD, tested to analyze dependency between attributes in L datasets, provides, in terms of accuracy, performances comparable with those obtained by traditional FAD, SVR and MP methods. On the other hand, a difficulty in using MFAD's for the analysis of massive datasets is the choice of the number of subsets: too low values of s can render the MFAD algorithm unusable as the subsets would remain massive and their size should be further reduced. However, too high values of s could affect the performance of the algorithm, resulting in fuzzy partitions too coarse grained or training sets not sufficiently dense compared to fuzzy partition.

A solution for choosing a suitable value of the parameter s is to check, in a preprocessing phase, if, by choosing a value for s, MFAD provides, on one of the s subsets chosen randomly, acceptable performance in terms of accuracy, i.e., an index of determinacy above the predetermined threshold. If this does not happen, then it is necessary to assign a greater value to the parameter s and repeat this process until determining a value of s that provides acceptable accuracy.

12.2.2   Fuzzy Transform for Massive Datasets: Future Perspectives

Therefore MFAD, tested to analyze dependency between attributes in L datasets, provides, in terms of accuracy, performances comparable with those obtained by traditional FAD, SVR and MP methods.

Formally, the task of finding a suitable value of s can be structured in the following steps:

1. Initially, a high value of s is assigned;
2. one of the s subsets is randomly selected;
3. the threshold value of the index of determinacy is fixed;
4. the FAD algorithm is executed using as training set the selected subset;
5. if the dependency is found then the partition in s subset can be used applying the MFAD algorithm, else the s value is decreased, and the process goes back to step 4.

A special care is needed in choosing the index of determinacy threshold α. The choice of a too high threshold could cause, in the preprocessing phase, the choice of a value of the parameter s too low, or the impossibility of using the MFAD algorithm as the subsets would remain massive datasets that cannot be handled by MFAD.

In addition, MFAD could be further improved by adopting strategies that avoid the overfitting problem. A K-folds cross-validation technique as the one discussed in Chap. 11 can be implemented partitioning each subset in K folds.

Of course, if K-fold cross-validation is applied to low dimension subsets, its effect of reducing the overfitting in the data is poor and the training data can result not sufficiently dense with respect to fuzzy partition; in this case it needs set a not too high value of the parameter s so that, by applying the K-fold cross-validation technique to each subset, the size of each training set is so as to respect the constraint of sufficient density with respect to the fuzzy partitions.

Applying the K fold cross-validation method to the MFAD algorithm the th subset DT_l with $l = 1,\ldots, s$, is partitioned in K folds $DT_{l1}, DT_{l2},\ldots, DT_{lK}$. In any cycle is constructed a training set $DT_{l1} \cup \ldots \cup DT_{lk-1}, \cup DT_{lk+1} \cup \ldots \cup DT_{lK}$ and a validation set DT_{lk} with $k = 1,\ldots, K$. We can measure two indices of determinacy by the formula (12.10): the index of determinacy r'^2_{clk} calculated in the training phase and the index of determinacy ρ'^2_{clk} calculated in the validation phase.

The results of the algorithm are given by aggregations of the results obtained applying the FAD algorithm in any cycle. For example, each component of the multi-dimensional direct F-transform can be calculated as the average of the values of this component calculated in each cycle, likewise, the training index of determinacy r'^2_{cl} and the validation index determinacy ρ'^2_{cl} are averages of the ones calculated in each cycle.

The Domain D_l, $l = 1,\ldots, s$, can be calculated by the intersection $D_{l1} \cap D_{l2} \cap \cdots \cap D_{lK}$ of the Cartesian product of the domains of the input variables obtained in any cycle.

Below we schematize in pseudocode the K-folds cross-validation MFAD algorithm.

<table>
<tr><td colspan="2">Algorithm: K-folds cross validation MFAD</td></tr>
<tr><td>Input:</td><td>Massive dataset DT composed by m instances with attributes $X_1, X_2,..., X_k, X_z$
Number of subsets s
Number of folds K</td></tr>
<tr><td>Output:</td><td>Direct F-transform components found for each subset
Domains $D_1,..., D_s$
Indexes of determinacy $r'^2_{c1},..., r'^2_{cs}$</td></tr>
</table>

1	Partition randomly the dataset DT in s subsets $DT_1,..., DT_s$
2	*for* $l = 1$ **to** s
3	Partition the subset DT_l in K folds $DT_{l1} \, DT_{l2},..., DT_{lK}$
4	$r'^2_{cl} := 0$
5	*for* k = 1 **to** K
6	Construct the k*th* training set $DT_{l1} \cup ... \cup DT_{lk-1}, \cup DT_{lk+1} \cup ... \cup DT_{lK}$
7	Calculate $\mathbf{F}_{lk}, r'^2_{clk}, D_{lk}$ via FAD algorithm
8	Calculate ρ'^2_{clk}, using DT_k as validation set
9	*next* k
10	Calculate as $\mathbf{F}_l, r'^2_{cl}, D_l, \rho'^2_{cl}$ as aggregations of $\mathbf{F}_{lk}, r'^2_{clk}, D_{lk}, \rho'^2_{clk}$
11	Store $\mathbf{F}_l, r'^2_{cl}, D_l, \rho'^2_{cl}$
12	*next* l
13	*end*

The index of determinacy ρ'^2_{cl} represents the average accuracy measured in the validation phase using the lth subset. If the value of ρ'^2_{cl} is comparable with r'^2_{cl}, then there is no overfitting in the model using the lth training set.

It could be adopted instead of r'^2_{cl} as a weight to assign to the lth inverse F-transform $H^F_{n_{1l}n_{2l}...n_{kl}}$ the assess the value of the output variable x_z, given a point P with input variable coordinates $(x_1, x_2,..., x_k)$, replacing the (12.11) with the following formula:

$$w_l(x_1, x_2, \ldots, x_k) = \begin{cases} \rho'^2_{cl} \text{ if } (x_1, x_2, \ldots, x_k) \in D_l \\ 0 \quad \text{otherwise} \end{cases} \quad l = 1, \ldots, s \quad (12.14)$$

In this way, estimates affected by the possible presence of overfitting would be avoided.

The architecture of the MFAD algorithm could also be implemented in other machine learning algorithms using multi-dimensional F-transform, such as the F-transform prediction method described in Sect. 10.4 or the MFC algorithm discussed in Sect. 11.5.

Future developments could use F-transform in hybrid models that apply innovative machine learning frameworks, such as deep learning or quantum computing architectures, in Artificial Intelligence applications that make use of massive data.

References

1. Dean, J., & Ghemawat, S. (2004). MapReduce: Simplified Data Processing on Large Clusters. In *OSDI'04: Sixth Symposium on Operating System Design and Implementation* (pp. 137–150). San Francisco, CA.
2. Wang, J., Liu, W., Kumar, S., & Chang, S.-F. (2015). Learning to hash for indexing big data: A survey. *Proceedings of the IEEE, 104*(1), 34–57. https://doi.org/10.1109/JPROC.2015. 2487976.
3. Laney, D. (2001). 3D data management: controlling data volume, velocity and variety. *Application Delivery Strategies Meta Group, 949,* 4.
4. Zikopoulos, P., Eaton, C., De Roos, D., Deutsch, T., & Lapis G. (2011). *Understanding Big Data: Analytics for Enterprise Class Hadoop and Streaming Data* (p. 166). McGraw Hill Professional. ISBN: 978-0-07-179053-6.
5. Chen, C. L. P., & Zhang, C. Y. (2014). Data-intensive applications, challenges, techniques and technologies: a survey on Big Data. *Information Sciences, 275,* 314–347.
6. Kitchin, R., & McArdle, G. (2016). What makes Big Data, Big Data? Exploring the ontological characteristics of 26 datasets. *Big Data & Society, 3*(1). https://doi.org/10.1177/ 2053951716631130.
7. Huber, P. (1997). Massive data sets workshop: The morning after. In J. K. Washington & D. Pregibon (Eds.), *Massive Data Sets, Proceedings of a Workshop* (pp. 169–184). Washington DC: National Academy Press.
8. Hathaway, R., & Bezdek, J. (2006). Extending fuzzy and probabilistic clustering to very large data sets. *Computational Statistics & Data Analysis, 51,* 215–234.
9. Wittek, P., (2014). *Quantum Machine Learning. What Quantum Computing Means to Data Mining* (p. 219). Academic Press. ISBN: 9780128009536.
10. Schuld, M., & Petruccione, F. (2014). An introduction to quantum machine learning. *Contemporary Physics, 56*(2), 172–185.
11. Schuld, M., & Petruccione, F. (2018). Supervised learning with quant.um computers. *Quantum Science and Technology, 287.* https://doi.org/10.1007/978-3-319-96424-9. ISBN: 978-3-319-96423-2.
12. Deng, L., & Yu, D. (2014). Deep learning: methods and applications. *Foundations and Trends in Signal Processing, 7*(3–4), 1–199. https://doi.org/10.1561/2000000039.
13. Lu, J., & LiBias, D. (2013). Correction in a small sample from big data. *IEEE Transactions on Knowledge and Data Engineering, 25*(11), 2658–2663.
14. Kim, J. K., & Wang, Z. (2019). Sampling techniques for big data analysis. *International Statistical Review, 87*(S1), S177–S191.
15. Jun, S., Lee, S. J., & Ryu, Y. B. (2015). A divided regression analysis for big data. *International Journal of Software Engineering and Its Applications, 9*(5), 21–32.
16. Perfilieva, I., Novàk, V., & Dvoràk, A. (2008). Fuzzy transforms in the analysis of data. *International Journal of Approximate Reasoning, 48,* 36–46.
17. Di Martino, F., Loia, V., & Sessa, S. (2010). Fuzzy transforms for compression and decompression of color videos. *Information Sciences, 180,* 3914–3931.
18. Segata, N., & Blanzieri, E. (2009). Fast local support vector machines for large datasets. In P. Perner (Eds.), *Machine Learning and Data Mining in Pattern Recognition. MLDM 2009. Lecture Notes in Computer Science,* (Vol. 5632, pp. 295–310). Berlin, Heidelberg: Springer. https://doi.org/10.1007/978-3-642-03070-3_22.
19. Cheng, C. H., Tan, P., & Jin, R. (2010). Efficient algorithm for localized support vector machine. *IEEE Transactions on Knowledge and Data Engineering, 22*(4), 537–549.
20. Zheng, J., Shen, F., Fan, H., & Zhao, J. (2013). An online incremental learning support vector machine for large-scale data. *Neural Computing Applications, 22*(5), 1023–1035.
21. Peng H., Choi D., & Liang C. (2013). *Evaluating parallel logistic regression models.* In *2013 IEEE International Conference on Big Data.* Silicon Valley, CA, USA, 6–9 Oct 2013. https:// doi.org/10.1109/bigdata.2013.6691743.

22. Huang, G.-B., Wang, D. H., & Lan, Y. (2011). Extreme learning machines: A survey. *International Journal of Machine Learning and Cybernetics, 2*(2), 107–122.
23. He, Q., Shang, T., Zhuang, F., & Shi, Z. (2013). Parallel extreme learning machine for regression based on MapReduce. *Neurocomputing, 102,* 52–58.
24. Chen, C., Li, K., Duan, M., & Li, K. (2017). Chapter 6: Extreme learning machine and its applications in big data processing. In *Big Data Analytics for Sensor-Network Collected Intelligence* (pp. 117–150). Intelligent Data-Centric Systems. https://doi.org/10.1016/B978-0-12-809393-1.00006-4.
25. Yao, L., & Ge, Z. (2019). Distributed parallel deep learning of Hierarchical Extreme Learning Machine for multimode quality prediction with big process data. *Engineering Applications of Artificial Intelligence, 81,* 450–465.
26. Di Martino, F., & Sessa, S. (2020). *Attribute Dependency Data Analysis For Massive Datasets By Fuzzy Transforms*. Soft Computing (in press).

Bibliography

1. Cox, I.J., Miller, M., Bloom, J., Fridrich, J., & Kalker, T. (2007). *Digital Watermarking and Steganography* (2nd edn., 624 pp). Morgan Kaufmann, San Francisco, CA, 200. ISBN: 9780123725851.
2. Dimitoglou, G., Adams, J.A., & Jim, C.M. (2013). Comparison of the C4.5 and a Naïve Bayes classifier for the prediction of lung cancer survivability. *International Journal of Computer Science and Applications, 6*(2), 256–261.
3. Hurrell, J.W. Decadal trends in the North Atlantic Oscillation: Regional temperatures and precipitation. *Science, 269*, 676–679.
4. Haralick, M., Perfilieva, I., & Hod'áková, P. (2012). Advanced F-transform-based image fusion. *Advances in Fuzzy Systems, 2012*. Article ID 125086, 9 pp.
5. Kitchin, R., McArdle, G. (2016). What makes Big Data, Big Data? Exploring the ontological characteristics of 26 datasets. *Big Data & Society, 3*(1), https://doi.org/10.1177/2053951716631130.
6. Klement, E.P., Mesiar, R., & Pap, E. (2000). *Triangular Norms* (387 pp). Dordrecht: Kluwer. ISBN 0-7923-6416-3.
7. Novak, V., Perfilieva, I., Romanov, A., & Yarushkina N. (2010). Forecasting trend of a group of time series using F-transform and fuzzy natural logic. In *The 11th International FLINS Conference (FLINS 2014)* (15 pp),·Sept 2014. https://doi.org/10.1142/9789814619998_0026.
8. Pal, S.K., & Mitra, S. Multilayer perceptron, fuzzy sets, and classification. *IEEE Transactions on Neural Networks, 3*(5), 683–697.
9. Perfilieva, I., Yarushkina N., & Afanasieva T. (2010a). Time series analysis by discrete F-transform. In *Conference Paper in IEEE International Conference on Fuzzy Systems*. https://doi.org/10.1109/FUZZY.2010.5584264.
10. Schuld, M., & Petruccione, F. (2018). *Supervised Learning with Quantum Computers. Quantum Science and Technology* (287 pp). https://doi.org/10.1007/978-3-319-96424-9. ISBN 978-3-319-96423-2.
11. Shih, F.Y. (2007). *Digital Watermarking and Steganography: Fundamentals and Techniques* (2dn edn., 270 pp). Boca Raton, FL: CRC Press. ISBN: 9781498738767.

© Springer Nature Switzerland AG 2020
F. Di Martino and S. Sessa, *Fuzzy Transforms for Image Processing and Data Analysis*,
https://doi.org/10.1007/978-3-030-44613-0

Index

A

Active systems, 49
Artificial Neural Network (ANN), 177, 194
Attribute dependency, 129, 197, 203
Autofocus, 49, 50, 54, 103
Auto-regressive Integrated Moving Average
(ARIMA), 138, 140, 141, 154, 162,
163, 167, 168, 170, 171

B

Bi-dimensional F-transform, 27, 64, 86, 95,
123
Bilinear Fuzzy Relation Equations (BFRE),
112, 115, 117, 120
Block
decomposition, 29–31
Block-wise algorithm, 103, 104

C

Canny detector, 74, 75, 77
Chaotic Darwinian PSO (CDPSO), 87, 88
Classification, 82, 173–179, 181–183, 188,
189, 193–195
Cluster, 54–57, 81, 82, 89–92, 97, 104, 105
Coding images, 27, 36
Color image compression, 36
Complete image fusion Algorithm (CA),
64–66, 68, 71, 72
Compression rate, 29, 31, 34–36, 41, 42,
44, 45, 72, 86, 95–98, 108, 109, 111,
115–117, 120
Contrast detection, 49, 50
Contrast Measurement Adaptive to Noise
(CMAN), 53, 58

Controlled Complete Algorithm (CCA), 68,
69, 71, 72
Crisp set, 1, 2, 8

D

Darwinian PSO (DPSO), 86–88
Data analysis, 123, 126, 137, 145, 195
Data classification, 173, 177, 179, 183, 193
Data regression, 195
Decision tree, 174, 176, 189
Decoding images, 27, 28, 32, 36, 123
Delta frame (Δ-frame), 41, 44–47
Direct fuzzy transform, 54, 129, 145, 183,
197, 199
Direct multi-dimensional fuzzy transform,
183, 186, 187
Discrete Cosine Transform (DCT), 30–33,
36, 54
Discrete fuzzy transform, 12

E

Edge detection, 61, 73–75
Energy of Laplacian, 52
Euclidean distance, 89, 90
Extended Simple Algorithm (ESA), 65, 67,
69, 71, 72

F

F^1-Fuzzy transform (F^1-transform), 23–25,
32–35, 74, 75, 77, 153, 155, 164–166
Forecasting, 137, 138, 140, 141, 144–146,
153, 157, 159, 163–165, 170, 171
Fragile image watermarking, 59, 105, 120
Frequency histogram, 54, 87, 175

© Springer Nature Switzerland AG 2020
F. Di Martino and S. Sessa, *Fuzzy Transforms for Image Processing and Data Analysis*,
https://doi.org/10.1007/978-3-030-44613-0

Printed in the United States
by Baker & Taylor Publisher Services